Resource Management in Advanced Wireless Networks

Scrivener Publishing
100 Cummings Center, Suite 541J
Beverly, MA 01915-6106

Publishers at Scrivener
Martin Scrivener (martin@scrivenerpublishing.com)
Phillip Carmical (pcarmical@scrivenerpublishing.com)

Resource Management in Advanced Wireless Networks

Edited by

A. Suresh

J. Ramkumar

M. Baskar

and

Ali Kashif Bashir

Scrivener Publishing

WILEY

This edition first published 2023 by John Wiley & Sons, Inc., 111 River Street, Hoboken, NJ 07030, USA and Scrivener Publishing LLC, 100 Cummings Center, Suite 541J, Beverly, MA 01915, USA

For more information about Scrivener publications please visit www.scrivenerpublishing.com.

Wiley Global Headquarters
111 River Street, Hoboken, NJ 07030, USA

For details of our global editorial offices, customer services, and more information about Wiley products visit us at www.wiley.com.

Library of Congress Cataloging-in-Publication Data

ISBN 9781119827498

Front cover images supplied by Pixabay.com
Cover design by Russell Richardson

Set in size of 11pt and Minion Pro by Manila Typesetting Company, Makati, Philippines

Printed in the USA

10 9 8 7 6 5 4 3 2 1

Contents

Preface

Many universities and networking organizations are working on wireless mobile networks. Many universities are offering some research based advanced topics courses like broadband wireless networks, wireless mesh networks, network coding, etc. and especially for the researcher. This new work describes the introduction of wireless networks, such as WiMAX, LTE, and 5G technology, based on scheduling, resource allocation and channel utilization. It identifies the existing challenges and its implications which pave the way to find solutions. The work provides possible solutions for resource allocation in wireless networks.

The book discusses the challenges that are emerging while managing the resources in various wireless networking technologies. Initially, the evolution of wireless networking technologies is deliberated with lot of advantages in improving the data rate and data reliability. Then the various architecture design is structured based on the network paradigm along with the evolution of networks based on the telecommunication industry. Then various salient features are highlighted in managing the resource and the role of routing strategies being addressed for real-time applications. Based on the routing mechanism, the problem of interference is addressed based on the network while integrating the learning algorithm. Then the factor of QoS is addressed with 5G networks and addressed based on healthcare and how they are assisted with cloud data.

Other than healthcare, challenges in sustainable networks, financial sectors, are discussed. Thus, resources being utilized in the various advanced networks is managed and their challenges are addressed.

Chapter 1: **Introduction to Next Generation Networks 5G and Beyond**
This first chapter discusses the general introduction on the next generation network on 4G and beyond 5G and reflects the development of the network under 3GPP standards. There are lot more benefits, which are associatedwith 5G networks as it enhances the speed on the communication linkgets deployed on the 5G networks. In 5G, various new technologies

getassociated in reliable data communication with very high speed with lowlatency. Moving on to 5G, various societal based safety and sustainabilityget ensured and it get transformed into industrial standards also. Implicithow 5G role into the environment will influence individual and industries? Finally, how 5G help to reconstruct secure and viable environment for the current and future community.

Chapter 2: **Architecture and Future Trends on Next Generation Networks** This chapter discusses the structural design on the evolution of network paradigm and its subsequent development. The evolution of the network moves drastically based on certain metrics, such as power consumption and energy efficiency. The network development provides various worthwhile landscapes for both real time and non-real time. Regarding the phases on the network and its services, it converges on managing the IP, integrating the network, extending the services. Finally, based on the architecture there are various unique requirements and features that are involved in order to attain possible benefits.

Chapter 3: **Evolution of Next Generation Networks and Its Contribution Towards Industry 5.0** takes up the phases and accomplishments involved during the evolution on networks towards 5G as it influence the telecommunication industry significantly. The significant development in the networking era drives towards the 5G and beyond based on the assimilation of Artificial Intelligence (AI), Internet of Things (IoT) and Block chain, etc. If it is beyond 5G, it offers managing the services on offer terrestrial, maritime networks, and airborne by taking up the larger devices with complex management capabilities. As there is a space to go beyond the 5G to take up the development on certain parameters such as, data rate, multiplexing, frequency range, type of switching, providing vast services as it earns both advantages and disadvantages. Finally, it paves the way for moving towards next generation of 7G as it takes up the well-defined communication on satellite with enhanced security.

Chapter 4: **Understanding the Salient Features Related To Resource Management in Broadband Wireless Networks** covers certain salient features geared towards managing the resource involved in network technologies based on the increase in the number of users and no compromise on the network traffic/data reliability based on services. Various characteristics related to radio channel is discussed as it results in providing the effectiveness in the traffic management and QoS. Th problem of resource scarcity is considered during the utilization process based on certain techniques related to Adaptive Modulation and Coding (AMC); Spatial Multiplexing and Multi Accessing technique. Then performance metrics on QoS is discussed in terms of IP mobility, Security and mobility Support to provide service and data flexibility.

Chapter 5: **Network Routing and Its Real-Time Practice in Broadband Wireless Networks** discusses the routing strategies in wireless broadband networks with perspective of practical real time applications. Narrate the wireless technologies community with lot more demand in terms of services to connect the peoples together via internet. First discussion on the types of broadband wireless networks such as, fixed network and dynamic nature on WiMAX technologies etc. Major issues related to security, which is inbuilt into the routing mechanism takes up various attacks involved as DoS, DDoS, Rogue and selfish backbone devices, etc. Finally, there is a huge space on security metrics towards providing the overall protection on the network based on authentication, access control, and encryption characteristics involved.

Chapter 6: **Routing Mechanism in Broadband Wireless Network** discusses the broadband services on to the remote village as it gets operated from the major cities with operator industries. The broadband wireless networks are designed to provide certain features like, high speed Internet and data service. In order to consider the above features to achieve robustness and scalability, routing mechanism plays a vital role with low-cost routing path. Routing plays a significant role as it discuss the classification on routing protocols as static, dynamic and default. Routing mechanism is discussed based on the network topology of centralized and distributed structure. By considering the problem of convergence communication and multi-Hop dynamic routing, various issues related to routing is addressed based on the existing survey.

Chapter 7: **Interference Problem in 5G with Radio Access Network** addresses the significant approaches of Radio Access Networks (RAN) as it makes certain system with the context of visual interface management, robust traffic control and network share management. By considering the performance of the system, managing the interference problem is the challenging task in the mobile networks. Based on the interference, multiple methods related to radio nodes are addressed to RAN. Thus, various salient features are addressed related to 5G Random Race Machine system includes seamless management, static traffics control, powerful bit-slicing communication control, and 5-generation loose operation and loss transfer education.

Chapter 8: **Interference Techniques Based on Deep Learning in Wireless Networks**: Standardized wireless networks on small scale is discussed as it states the faster delivery along with improved execution time in order to better the performance of the network. To take the network performance to the next level, frequency band has to get enhanced as it results in frequency congestion as it is the serious problem in wireless communication. The problem of interference occurs during the enhanced network configuration. Thus, interference discussion makes the learning-based algorithm to involve the real time applications with resource allocation, traffic flow prediction, and data identification.

Chapter 9: **Implementing SDN Process-Based MRMC Wireless Networks**: Various quality service factors based on bandwidth, packet loss, Quality of Service (QoS), etc., on Multi-Radio Multi-Channel (MRMC) network are addressed in order to process the QoS on all routers being used. Various QoS challenges are highlighted based on some key metrics such as, limitations in the bandwidth, redundancy removal, limiting the buffer size and trade off based on energy/delay. Based on routing concept, Software Defined Network takes the QoS enhancement on wireless sensor networks. Based on the QoS discussion, various performance evaluations are performed based on modeling the network, design load and configuring the network simulation.

Chapter 10: **Advanced Wireless Mobile Network on Financial Literacy**: There is a drastic development in the wireless network as it results in the financial development and literacy rate. Based on the financial characteristics, there will be more improvement in the financial orders among the peoples as they are moving towards the digital era. There are 3 objectives are addressed as it makes more attentive towards the financial attention, finding to save the habits of the people and give valuable suggestion to improve the peoples habitation. Then various literature surveys is carried out to investigate the financial and literacy factors in various countries. Thus, the literacy removal is related with the advancement in the wireless networks is discussed.

Chapter 11: **Advanced 5G Technology in Healthcare Field**: In the recent digitalized era, enormous amounts of wearable devices are used in our bodies. The requirement of the monitoring the persona health and identifying/preventing the diseases is needed seriously as it occurs frequently among the old age persons. Regarding the solutions to various problems on wearable healthcare applications, there is a lack in providing the medical attention among the patients and continuous tracking on the person's healthcare. Thus, due to the significant growth in the wearable devices helps to provide vast research interest in the field of communication and telemedicine.

Chapter 12: **Cloud Assisted Data Storage in e-Healthcare System**: Healthcare system plays a vital role in the current era as it deals with vast amount of information. Based on the development in the storage address of Version 6 and cloud repository, the users and admin can stores all kinds of information including confidential and private data. Maintaining the data privacy plays a significant role as the data is being shared among the known and unknown users. Then discussing the encryption based section data techniques in order to ensure the privacy among the data and secure them. Suitable algorithm is identified and compared with other existing algorithms to perform better based on healthcare data repository. To evaluate the system performance, certain metrics are used such as, data precision, recall, F-measure and data accuracy.

Chapter 13: **Challenges in 5G Building Sustainable Networks and Its Opportunities**: 5G technology is already being deploying in various developed countries as it gives severe environmental data acquisitions with negative impacts. General highlights of the 5G technologies makes to provide vast internet access with 1 Gbps as it is offered as public services. For urban and rural areas, the internet speed with at least 100 Mbps. Deploying the 5G based on green environment makes to enhance and conserve the energy, managing the transceiver, transmission based on radio green and establishing the renewable resource into the framework. Thus, makes green based network deployment makes the world to be more better perfect.

Chapter 14: **Enhanced Human Growth Through Technical Financial Literacy**: The enormous growth of education, the standard of living, knowledge etc., makes significant knowledge on financial literacy very important. As it is the initial step for the growth for literacy as it gives massive growth on the education. In the developing and under developing countries, poverty problems takes challenging issue as there is no formal sources for the requirement on literacy. The main objective of this study is to make attention among various peoples of under development and save the peoples habitats. Based on the convenience sampling method, input data is collected to construct the financial literacy. In order to recover the financial literacy level, digitalizing the awareness based on cloud, wireless networks, etc.

The editors thank the contributors most profoundly for their time and effort.

Dr. A. Suresh
Dr. J. Ramkumar
Dr. M. Baskar
Dr. Ali Kashif Bashir

1

Introduction to Next Generation Networks 5G and Beyond

B. Vidhya*, H. Harshene and N. Abimathi

Department of Biomedical Engineering, Dr. N. G. P Institute of Technology, Coimbatore, Tamil Nadu, India

Abstract

The fifth generation of mobile communication technology is the near future. There's a lot of ambiguity about the arrival of 5G — the next generation of mobile network technology with speeds at least ten times faster than today's 4G networks. The gradual transition to 5G from earlier versions is intended to provide users with a more responsive and prospective services. It appears predestined to open up previously unimagined possibilities for innovation and growth in the domains where huge data are required such as Tele health, Virtual Reality, artificial intelligence, smart manufacturing and so on virtual health care. This chapter - 5G network and beyond - provides an overview of the benefits of 5G, as well as its evolution, constraints, and much more. To comprehend the rapidly approaching 5G, a thorough understanding of the preceding versions right from 1G to 4G is required which will provide you a broad vision of why we need to upgrade to the next generation. Misinterpretations of 5G are also examined in order to gain a clear understanding of what 5G is, what it is capable of, and its exclusivity. This chapter concludes with the effects on consumers and organizations in everyday life. It will also clarify the new equipment needed for a 5G connection. The wireless technology that is fast approaching is the fifth generation is on its way to usher a new age of greater, all-encompassing wireless broadband experiences for nearly every sector, product, and connectivity.

***Keywords*:** 5G, evolution, mobile communication, 3rd generation partnership project, working groups

**Corresponding author*: vidhyamkumar@yahoo.com

A. Suresh, J. Ramkumar, M. Baskar and Ali Kashif Bashir (eds.) Resource Management in Advanced Wireless Networks, (1–18) © 2023 Scrivener Publishing LLC

1.1 Introduction

The mobile phone technology landscape has changed quite a bit from the first generation over the last few years. During the first generation, phones were small enough to fit in a briefcase and there were many short conversations between most professionals. In the lead as much as 2G, the call for cellular offerings grew and in no way slowed down. Phones that are good enough to fit in your pocket, SMS and the internet are the indicators of the 3G world. And 4G brought smartphones, Youtube, and various browsers. Now, 5G absolutely remolds all our expert and private lives via permitting new use instances like Augmented Reality and more suitable games.

The 5G network is the fifth generation of mobile networks, with theoretical peak speeds of 20 Gbps compared to 1 Gbps for 4G. Voice and text were enabled in the first two generations, and 3G delineated the move to broadband access by supporting data rates of hundreds of kilobits per second. The current 4G industry is transitioning to 5G, with a tenfold increase in data rates approximately.

Cellular technologies are now focusing on providing a guaranteed connection. The fifth generation networks are aiming to uplift the connectivity to a higher level by utilizing cloud-based services to the users. This cloud service will help for the virtualization of 5G that will operate on software.

As a result of the 5G network, users will be able to move seamlessly between Wi-Fi and cellular coverage without requiring user intervention or any in-depth configuration. 5G networks will facilitate the portable customers to stay connected even during their movement from open air to closed infrastructures remotely. This connection will be established without the involvement or validation of the customers.

Mobile telephony is imminent for the 5G systems, which can also be termed as IMT-2020 (International Mobile Telecommunication). The term "5G" includes not just mobile cellular systems, but also Internet of Things (IoT), satellite communication systems, vehicle-to-everything (V2X) communications, and wireless local area networks (WLANs).

1.1.1 Benefits of 5G Networks

The ultimate motive of the upcoming generation of communication is to have better efficiency than the existing generations. Some of the advantages are as follows

- Enhancement of downlink and uplink communication speeds to 20 and 10 Gb/s, respectively
- The network capacity of the 5G network is being increased by ten thousand times than the existing networks
- An increase in data volume of 1000-fold per geographical area
- The ability of Wireless Sensor Networks (WSN) to accommodate hundred thousands of connections at once
- In comparison with the fourth generation, 5G exhibited increased spectral performance
    - Downlink - 30bit/s/Hz
    - Uplink - 15 bit/s/Hz
- Appreciably 5G network provides a reduction of around one ms latency time
- Better velocity restriction of 500 km/h is obtained for customers
- Spatial variety is improved
- Enhanced signaling efficacy

1.2 5G Evolution

Regardless of customization & popping up of new technologies, broadcasting and networking are among the foundations of human evolution and progress in technology. Presuming the absence of the various modes of transmission of data & communication, People would have been struggling in a generation with limited technology.

Mobile phones and other handheld devices rely on wireless communication that has grown over a few generations. The first generation is the first voice-only network and now it is advancing to 4G and eventually 5G. What are the essential ideas behind these wireless communication technologies and what are the changes happening in reality?

The term G in networks represents the generation. The technology innovation for every generation is represented by a number along with G. For example, 3G represents the third generation. The whole world is widely using 4G networks now and we are marching towards 5G. Before understanding 5G, it is mandatory to know about the previous generations.

According to the researchers, new generations in the cellular mobile system have grown drastically to a higher level and new generations are introduced every ten years since the introduction of the first generation (1G) in 1981. A technological revolution happened due to the mobile industry

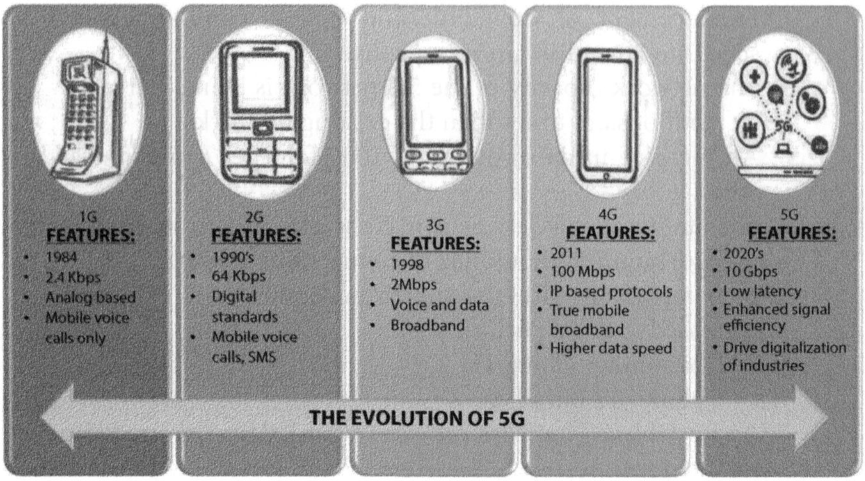

Figure 1.1 The evolution of 5G.

through the advancement of networks from 1G through 5G over three decades. We must thank 1G because it is the reason for the stepping stone for an abundance of mobile telephony retail market. 2G guaranteed universal information sharing and reliable cellular networks, as well as Short Message Service (text messaging) in this revolution. 3G furnished faster communication. 4G contributed towards an increased speed and communication data which enabled the customers to surf the internet with ease. With remarkable data capabilities, boundless call capacities, and inestimable data transmission, 5G technology will be the potent cellular wireless network in the near future (see Figure 1.1).

1.2.1 1G-Analog Cellular Networks

The 1G network was built for the purpose of voice-only communication that worked with the speed of 2400 bits per sec and analog in nature. After the implementation of the 2G networks only, the 1G network was named. The first generation utilized a number of cellular nodes to connect a call. When the customers move from one place to another during the call, cell nodes were used to keep the call connected. Feeble connection for voice communication, reduced capacity and unpredictable handoff are just a few of the drawbacks of 1G. The early phones, which used analog technology, were huge and bulky. Radio towers were used to replay the audio communications in 1G which in turn paved the path for the intruders to overhear the conversation. This was a vulnerable act that could not be tolerated.

1.2.2 2G - Digital Mobile Networks

The 2G (second generation) wireless networks were instigated in the early 1990s, which were based on digital rather than analog standards. Rapid phone-to-network signaling was facilitated by 2G digital networks, which aided the introduction of mobile phones that can be used through the prepaid plan of the network provider. The second generation facilitated people with Short Message Service (SMS) with the help of Global System for Mobile Communications (GSM) initially, and over other digital networks later. 2G networks provide wider benefits such as lower battery consumption, improved voice clarity, and less line noise. Data and voice calls were kept private and secure thanks to digital encryption. Finally, digital transmissions are thought to be eco-friendly.

People used 2G networks during the year from 1980 to 2003 which had many developments in a spectrum like

GSM: With GSM (Global Systems for Mobile Communication), audio and video can be transmitted at 30-35 kilobits per second speed. It had a huge impact on the progress of mobile technology, as it was implemented at a time of booming mobile phone usage.

GPRS: With (General Packet Radio Service) GPRS, data transmission occurred at a faster speed of 110 kbps than GSM.

EDGE: In the year 2003, EDGE (Enhanced Data for Global Evolution) was launched and provided numerous improvements over GPRS and GSM. 2.9G or 3G are the other names given to the EDGE. Even now, various internet service providers use this technology globally as the base one because all the customers are satisfied with the service provided by it at the speed of 135 kbps.

1.2.3 3G-High-Speed Data Networks

Researchers developed the third-generation (3G) networks in the year 1998 after doing certain improvements over 2G networks. The aim of this 3G wireless technology is to increase the rate of data transmission and practically the uploading and downloading speed was provided at a 2 Mbps rate. Customers were able to communicate emails, images, audio and video files at an express rate even when the files are huge in size. People were happy as they were able to view movies, live sports or videos online without buffering time as the streaming was very fast in 3G. They experienced high security online and they started enjoying surfing the internet often. The technologies that were extensively employed in 3G are Enhanced Data for Global Evolution and Code-division multiple access 2000.

In 3G technologies, long distance communication was made easy and simple and signal transmission for it was established with reliable connection by means of Smartphone towers. The packet switching was used in 3G rather than circuit switching for data transmission which made a tremendous increase in efficiency than 2G networks. By using 3G's high-speed connection, users were able to stream radio and TV programs to their handsets. Video conferencing and quicker web surfing had been additionally viable due to the technology.

In a recent survey by the telecom companies, it was found that the download speeds have been increased to 7.2 Mbps and upload speed to 2 Mbps. In July 2000, the HSPA (High-Speed Packet Access) technique was incorporated in 3G to show improvements in the protocol operation which was referred to as turbo 3G, 3 G+ or 3.5 G. of mobile telecommunications protocols included a new enhanced 3G mobile telephony protocol called 3.5G, 3G +, or turbo 3G. If turbo 3G+ works on UMTS networks, then the users experienced a higher speed of communication with a few clicks.

1.2.4 4G Expansion of Mobile Broadband

In 2011, Verizon launched the country's first 4G cellular network that offered ten times better speed than 3G networks. 4G connections were able to provide an expeditious speed of 14Mbps to 150Mbps for downloading purposes that employed Internet protocol-based networks. Even voice data is transmitted over IP. 4G utilizes a packet switching network as a communications protocol for data transmission. Due to these standardized packets, 4G allows data to pass through all kinds of networks without being disarranged or distorted.

4G network is an advancement of the 3G network which provides greater capacity, solutions, and exquisite streaming of audio/video files. 4G delivers a significant improvement in data quality and efficiency over 3G networks due to the reason that the data rate has been increased to one hundred Mbps. Circuit switching is eliminated in 4G networks to support every Internet Protocol network. Users were able to access the web and watch high-definition movies and videos on their mobiles due to the high-speed 4G network. The 4G network was liked by users across the globe since the huge volume of data can be downloaded in a few seconds which saved their time and provided the comfort zone of technology. As a result of technical breakthroughs, smartphones have grown into modern systems. 4G is also known as MAGIC because it flourished **M**obile **A**nytime **G**lobal **I**ntegrated **C**onnection.

4G wireless network was elevated further to contribute to very high speed data rate using the Long Term Evolution (LTE). In this elevation, the 3G framework was fully altered and standardized to lower transmission delay, better performance and response times.

Due to various improvements that took place in the area of wireless systems with respect to general growth and tremendous alterations in the fundamental operations over these decades, people are enjoying the modern era of technology developments.

1.2.5 5G—Design Innovation

Apart from the previous 4G LTE network, the upcoming remarkable state of mobile networks is the fifth generation network with 1-10Gbps speed. 5G networks will be used globally sooner in 2022. 5G technology will offer incredible data bandwidth and unrestricted data transmission. Faster-calling rates, multi device connectivity, lower power requirements, simultaneous redundancy, increased mobile transmission, reduced variable latency, higher capacities and reliable tool-to-tool interactions are a few of the primary aspects of 5G networks.

1.3 5G - Exclusiveness

5G operates on the identical radio frequencies which are being used in mobile phones, but the technology can be utilized much further.

Apart from its ability to download a high-quality movie or picture in a few seconds, 5G aims to connect everything throughout the world - certainly, and without latency - in order to make people monitor, analyze and control everything in reality.

1.4 Challenges of Networks

Networks starting from 1G to 4G have certain constraints of their own. Each evaluation of every generation provides solutions for the previous generations. But consistently, certain challenges still follow as a shadow but the intensity of its impact varies. The general challenges of the networks are listed in Table 1.1 with their limitations. All the limitations and the

Table 1.1 General challenges of networks.

Challenges	Mitigation approach	Constraints
Pilot contamination	Chain network uses pilot sequence division, zero-forcing (ZF) precoding	Band FD M-MIMO uses pilot sequence, non-reciprocal transceiver
Drones	ICIC design UAV UL, multi-beam UAV UL, Echo state network	Downstream intervention difficulties for cellular users and UAVs in various circumstances
Finding Base Station	Spectrum pooling coordination, pencil beam formation, Interference avoidance	Periodic packet loss, latency, the impact of real world execution
Control channel	Sparse power control, power level	Joint development of control and data channels, coverage gaps
Receiver configuration	Non-linear processing	Spacing of antenna element

challenges till 4G are addressed by incorporating new technologies in 5G, a reliable low latency network.

1.5 What Will 5G Accomplish?

Besides improving network connections dramatically, 5G provides us with unprecedented opportunities to deliver stunning solutions that reach across all aspects of our lives.

In the future, billions of physical and digital devices will gather and share real-time information to prevent road accidents; or lag-free guaranteed connections will launch life-saving applications into the air, or manufacturing lines sufficiently predictive that they can block interruptions far before they happen.

1.6 Progressing Societies

With 5G, safety and sustainability can be enhanced in groundbreaking ways.

- Perceptive power grids for significantly decreased carbon emissions.
- More linked vehicles sharing information to avoid collisions
- Quick promptness of emergency services in case of an accident.
- Natural disasters can be detected early with connected sensors.
- The emergency response becomes faster and more effective with drones.
- Experts from remote locations can consult and diagnose patients with ease.

1.7 Transforming Industries

- The future of business depends on 5G enabling flexibility, efficiency, and responsibility.
- Supply and demand are handled autonomously by production lines.
- Real-time failure of machines can be foreseen using digital copies.
- Networks that are used for logistic purposes provide direction to the goods cargo intelligently depending on constraints of the real world.
- Traceability of items all the way down to the individual item at warehouses and ports.
- Robots and vehicles that can operate remotely to ensure safety in dangerous environments.
- IoT is used in agriculture to increase efficiency in crop production.

1.8 The Invention of 5G

5G was not invented by any company or an individual because many mobile companies worked together to make this extraordinary networking exist in the world.

1.8.1 Technologies of 5G

The mobile ecosystem has several companies that are enabling 5G to exist, but no single company or person owns it. But few companies like Apple, Qualcomm, Ericson and so on claim credit for inventing many of the key technologies that form the 5G foundation in the uprising wireless communication generation.

The 3GPP is the third Generation Partnership Project is a work cluster that makes global specifications related to Universal Mobile Telecommunications System for 3G networks in addition to high-speed packet access, 4G Long-Term Evolution, and fifth generation networks, at its center.

3GPP utilizes various key improvements in all components of 5G architecture from connecting from the application layer of the source to the destination device. Other contributors of this technology show diversification from architecture operators and element manufacturers to mobile band users and internet service providers.

The 5G network is principally based on OFDM, which modulates digital data over numerous varying transmission mediums to lessen the intrusion of communication.

Higher bandwidth telecommunication technology like Sub-6GHz as well as millimeter wave is used in the 5G New Radio technique combined with OFDM principles in the 5G network. When the working procedures of the 5G with OFDM and 4G LTE are compared, we can say that both are the same. When OFDM is employed in the 5G NR wireless interface, adaptability and expandability are greatly increased which in turn empower access of numerous communities and objects effectively in all kinds of situations.

The 5G network utilizes the spectrum resources ranging from sub-3 GHz to 100 GHz in the transformation from 3G to 5G, resulting in greater bandwidths. 5G will be capable of operating under the two bands namely sub-6 GHz and millimeter-wave bands i.e. 24 GHz and higher, resulting in low latency, increased capacity, reduced latency with Gbps output and low latency.

5G is envisioned to broaden its application area in delivering a wide number of services like Internet of Things connectivity to a huge network and establishing an impossible connection into reality. To facilitate this, Time-division duplex design is used in numerous innovative 5G New Radio techniques.

In the end of the year 1998, 3GPP was launched by (ETSI) European Telecommunications Standards Institute in collaboration with other global SDOs for the development of technology innovations of the third generation.

Initially, 3GPP was impacted more by the existing 2GT DMA-based GSM standard. Simultaneously, the state of the art cluster established the 3rd Generation Partnership Project 2 (3GPP2) in the USA, with the goal of developing universal specifications for the third generation networks deployed on the progression of the second generation IS-95 CDMA principles. The enhancements in this technology were achieved by the active and continued evolution of new standards by the several businesses, including Intel, HiSilicon, Qualcomm and Apple though they are competitors of one another.

Although certain discrepancies remained, both 3GPP and 3GPP2 eventually narrowed down to use instigated CDMA technology which formed the foundation for the standards of 3G networks. W-CDMA or UMTS is the other name given to 3GPP3 technologies which used the carrier bandwidth of five MHz. CDMA2000 is the technology of 3GPP2 technologies which utilizes a bandwidth of 1.5 MHz for carriers.

The above two technologies are recognized as 3G standards by the expert organization for information and communication of the United Nations named International Telecommunication Union. It was further developed with innovative data optimization technologies (3GPP2 EVDO, 3GPP HSPA) that are used globally.

This gets the people the fifth generation of networks. This 5G is likely the very much determined generation yet, focusing not only to bring the latest extent of performance and competence to improve the broadband experiences and also to develop an integrated service for an ample of applications at an express rate of connectivity. To fulfill these requirements, in the year 2016, 3GPP started to do research and development in establishing the Universal standards for the fifth generation -> 5G NR (New Radio) -> 5G NextGen (Architecture for next generation). Simultaneously, 4G LTE continued to improve since 5G NextGen will become critical. ITU 5G/IMT 2020 standards work on the basis of joint collaboration of 4G LTE and 5G NR. The technologies used in 5G are considered to be unique and organizations compete for the development of 5G compared to the previous generations. People across the world know that the 5G will provide a massive impact on almost all forms of businesses and areas, so understanding the architecture and the working method of the 3GPP is vital in the technology world.

1.8.2 Misinterpretations of 3GPP -Addressed

- 3GPP evolves technical specifications, rather than standards this is a minor but critical clarification of companies. The 3GPP is indeed an engineering group that develops design details that are subsequently converted into standards by the SSO's (seven regional Standards Setting Organizations) that are responsible for designing and implementing an IPR strategy.
- 3GPP is an organization that has been guided by its members. R&D, industrial innovations and other alliances of individual members of the whole world will take care of all the technical projects in addition to the usual IT services offered by 3GPP. The chairpersons of the various 3GPP groups are elected amongst all the member companies across the globe and are expected to work harmoniously abiding to the objectives of 3GPP.
- 3GPP works on very minimal end-to-end oversight with a disjointed way of working. It specifies the requirements for a complete end-to-end communication system, comprising mobile devices, remote interface, Internet service providers and the network architecture. The intricacy and the expansion of such networks are huge, so in order to provide more sophisticated services like cyber security and remote connectivity, fragmentation of the various work areas must be increased.

As shown in Figure 1.2, 3GPP is grouped into sixteen distinct Working Groups (WG). These Working Groups, together with the three prevailing Technical Specification Groups (TSG), are responsible for the majority of the innovation work and decisions.

1.8.3 3GPP's Real-Time Progress Indicator

With the constant evolution of generations of cellular technologies every decade, new specifications and new services that are similar to releases of prominent OS for smartphones/PCs offer distinguished value to the ecosystem around us.

3GPP is wrapping up work on a lot of astonishing releases and a number of other launches simultaneously happen at varying phases while work is currently in progress as illustrated in Figure 1.3. All upgrades are fully

Radio Access Network (RAN) Technical Specification Group Describes the radio communications between UEs and core network		Service/System Aspects (SA) Technical Specification Group Responsible for overall architecture & service capabilities	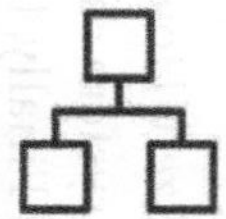	Core network & Terminals (CT) Technical Specification Group Responsible for core network; defines terminal interfaces & capabilities	
RAN WG1	Layer 1 (Physical) spec	**SA WG1**	Service requirements	**CT WG1**	Mobility Mgmt, Call Ctrl, Session Mgmt
RAN WG2	Layer 2 and 3 (RR) protocols	**SA WG2**	Architecture	**CT WG3**	Policy, QoS and Interworking
RAN WG3	Access network interfaces + O&M	**SA WG3**	Security	**CT WG4**	Network protocols
RAN WG4	Performance requirements	**SA WG4**	Codecs, multimedia system	**CT WG6**	Smart card application
RAN WG5	UE conformance testing	**SA WG5**	Telecom management		
RAN WG6	Legacy RAN, e.g. GSM, HSPA	**SA WG6**	Mission-critical services		

Figure 1.2 The structure of the 3GPP distributed organization

Figure 1.3 3GPP-flashcard.

fixed and available for execution when a launch is finalized. However, every 3GPP launch is independent, implying that a telecom system can be developed utilizing the pack of fixed criteria in that launch. Finally, launches do not simply include newly developed features; they're also extremely cyclic, based on previous launches.

The working procedures of the 3GPP and this generation network cannot be understood easily as the complication involved in this is higher because it was developed by collective determination amid hundreds of diverse entities with potentially disparate interests.

1.9 Role of 5G

The fifth generation is bringing a revolution in all the domains of the world through its technological development (see Figure 1.4).

1.9.1 Impact on an Individual

- 5G is intended to do wonders in human life through providing with greater browsing experience with faster uploading and downloading speeds, connecting a thousand million devices with ease mainly in the field of Internet of Things (IoT), connecting people anywhere, virtual reality (VR) and artificial intelligence (AI).

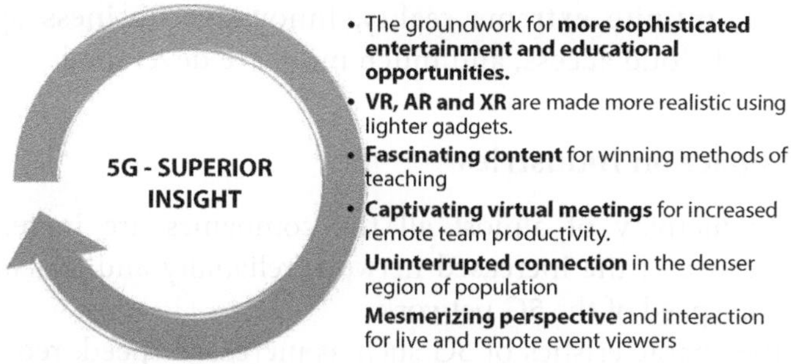

Figure 1.4 A glimpse of 5G services.

- To quote few, 5G will help to access improved and novel amenities like near-immediate cloud service at the fingertips, experiencing wonderful gaming scenarios for more than one player, exaggerated reality purchasing and live video collaboration.
- Furthermore, 5G will provide innovative alluring practices like virtual reality and Augmented reality with quicker, increased data rate, reduced time delay, and economical usage expenses in smartphones.
- 5G can be used for the industry transformation through its ultra-reliable networks with low latency in the services like controlling the critical infrastructure, automobiles and medical systems remotely.
- 5G connects an enormous number of embedded sensors remotely without a glitch with the property of giving incredibly lean and economical connection solutions with reduced power.
- Data Usage by an average individual through smartphones is predicted to be 12GB/month in the year 2022-2023. The reason behind this is that they are exposed to tons of sources of information in the form of entertainment and IoT and cloud computing services.
- 4G has dramatically revolutionized the way people use data. Industries developing mobile have grown to huge heights introducing bank transactions, video streaming, food delivery, transportation sharing, and much more.
- 5G will broaden the telecom community in all the domains of the modern world. Due to this, the development of cutting-edge user experiences such as continuous

IoT, infinite extreme reality, innovative business apps, rapid cloud access, and much more are developed.

1.9.2 Impact on Industries

- Productivity of almost all the companies are increased because of the increased network reliability and transmission speed of the 5G network.
- The characteristics of 5G such as increased speed, reduced latency and always connected network make certain organizations that are designed as smart factories to operate industrial Ethernet to improve operational productivity and precision effectively.
- 5G can be used in remote-controlled equipment that relies on time-critical communication to stop accidents and dangers.
- Mission-critical use cases can be achieved by the increased emphasis on availability, robustness, and resilience of 5G to serve the new business.
- Data protection is guaranteed with the confidential computing technologies of 5G.
- 5G can be used to build Smart cities to make the residents of that place live a high tech life with all kinds of regions like infrastructure, vehicle safety, VR, media and entertainment more connected at a faster rate under one umbrella at high efficiency.

1.10 Is a New Phone Required for 5G?

Yes, new smartphones are required to use the 5G network. Smartphones with Snapdragon 5G Mobile Platforms are capable of supporting 5G. Nowadays, to enable the usage of 5G by the customers, internet service providers across the globe are offering 5G services and mobile phone manufactures are introducing new models to support this enhanced service. Once 5G technology is implemented and its compatible smart devices become more widely available, we can expect the release of a variety of smartphones and carrier subscriptions at regular intervals of time.

1.11 Summary

The smartphone has become an integral part of our daily life. The average person spends about three hours every day on their smartphone for various purposes like downloading, uploading, sharing, interacting, browsing, and so on. The people started to adore the way mobile devices were capable of accessing high-speed Internet anywhere anytime. In order to achieve greater efficiency and throughput continuous evolution of the 3G & 4G Volte mobile technologies takes place regularly across the world.

5G – the next generation of wireless technology will bring a new era of superior, comprehensive wireless broadband experiences to almost every industry, product, and link. 5G provides faster connections than ever before while creating brand new opportunities for businesses and individuals.

Increasing connectivity speeds, reducing latency, and increasing bandwidth allow societies to advance, industries to transform, and to make every day enjoyable. Telehealth, connected automobiles, road/air regulatory systems, sophisticated gaming and video streaming with cloud services used to seem ultramodern, but now they're mainstream.

5G technology will build a brighter, secure, and highly viable future for the people. Let us join and embrace the latest technology and browse faster and stay connected with people, devices and the world and enjoy the benefits of this technology revolution.

Bibliography

1. M.-D. Kim, J. Liang, J. Lee, J. Park, and B. Park, "Directional multipath propagation characteristics based on 28GHz outdoor channel measurements," in 2016 10th European Conference on Antennas and Propagation (EuCAP), 2016, pp. 1–5. DOI: 10.1109/EuCAP.2016.7481755.
2. C. Ziółkowski and J. M. Kelner, "Antenna pattern in three-dimensional modelling of the arrival angle in simulation studies of wireless channels," IET Microw. Antennas Propag., vol. 11, no. 6, pp. 898–906, May 2017. DOI: 10.1049/iet-map.2016.0591.
3. 3GPP, "5G. Study on channel model for frequencies from 0.5 to 100 GHz (3GPP TR 38.901 version 15.0.0 Release 15)," European Telecommunications Standards Institute (ETSI), Sophia-Antipolis, France, ETSI TR 138 901 V15.0.0, Jul. 2018.
4. J. M. Kelner and C. Ziółkowski, "Multi-elliptical geometry of scatterers in modeling propagation effect at receiver," in Antennas and wave propagation, P. Pinho, Ed. London, UK: InTech, 2018, pp. 115–141. DOI: 10.5772/intechopen.75142.

5. ITU-R, "Recommendation ITU-R M.2083-0: IMT Vision – Framework and overall objectives of the future development of IMT for 2020 and beyond," International Telecommunication Union (ITU), Geneva, Switzerland, Rec. ITU-R M.2083-0, Sep. 2015.
6. M. Sauter, From GSM to LTE-Advanced Pro and 5G: An introduction to mobile networks and mobile broadband, 3rd ed. Hoboken, NJ, USA: Wiley, 2017.
7. Alex Mavromatis, Aloizio Pereira Da Silva, Koteswararao Kondepu, Dimitrios Gkounis, Reza Nejabati, Dimitra Simeonidou, "A Software Defined Device Provisioning Framework Facilitating Scalability in Internet of Things", 5G World Forum (5GWF) 2018 IEEE, pp. 446-451, 2018.
8. Abdul Ahad, Mohammad Tahir, Kok-Lim Alvin Yau, "5G-Based Smart Healthcare Network: Architecture Taxonomy Challenges and Future Research Directions", Access IEEE, vol. 7, pp. 100747-100762, 2019.
9. Mamoona Humayun, Nz Jhanjhi, Madallah Alruwaili, Sagaya Sabestinal Amalathas, Venki Balasubramanian, Buvana Selvaraj, "Privacy Protection and Energy Optimization for 5G-Aided Industrial Internet of Things", Access IEEE, vol. 8, pp. 183665-183677, 2020.
10. Cong Peng, Min Luo, Huaqun Wang, Muhammad Khurram Khan, Debiao He, "An Efficient Privacy-Preserving Aggregation Scheme for Multidimensional Data in IoT", Internet of Things Journal IEEE, vol. 9, no. 1, pp. 589-600, 2022.
11. Libin Zheng, Jing Chen, Tonglei Liu, Bingnan Liu, JiaNan Yuan, Ganghong Zhang, "A Cloud-Edge Collaboration Framework for Power Internet of Things Based on 5G networks", Information Communication and Networks (ICICN) 2021 IEEE 9th International Conference on, pp. 273-277, 2021.
12. Kinza Shafique, Bilal A. Khawaja, Farah Sabir, Sameer Qazi, Muhammad Mustaqim, "Internet of Things (IoT) for Next-Generation Smart Systems: A Review of Current Challenges Future Trends and Prospects for Emerging 5G-IoT Scenarios", Access IEEE, vol. 8, pp. 23022-23040, 2020.

2

Architecture and Future Trends on Next Generation Networks

M. Saravanan[1]*, J. Ajayan[2], Eswaran Parthasarathy[3] and V. Ramkumar[4]

[1]*Sri Eshwar College of Engineering, Electronics and Communication Engineering, Coimbatore, India*
[2]*SR University Warangal, Electronics and Communication Engineering, Telangana, India*
[3]*SRM Institute of Science and Technology, Electronics and Communication Engineering, Chennai, India*
[4]*Vellore Institute of Technology, Electronics and Communication Engineering, Tamilnadu, India*

Abstract

Many next-generation network solutions are focused on implementing innovative protocols to improve low power consumption and energy efficiency, because this will be the main challenge in all aspects of future networks. Energy-saving protocols and equipment on wired networks can reduce energy prices for service operators, data centers, and large organizations that rely heavily on a network technology to provide customer access. In wireless sensor networks, energy-saving protocols can extend the life of sensor networks, which is one of the biggest challenges in this field. Next-generation network architecture is low-latency, automated, easier to manage, easier to customize, confident, and high performing. It works closely with service providers and uses network analytics to manage traffic flow. Wireless has the benefit of now no longer in need of big infrastructure investments of stressed networks, which opens up the marketplace to smaller companies. This additionally makes it extra possible for huge corporations, in addition to customers, to exchange for wireless products. These approaches could help provide internet access to rural communities, remote communities, and people in developing countries. Access to the Internet could make a big difference for these groups as well as for those who live in more remote parts of the country in the developed world. These changes with the new improved feature results in the aspects of speed, dependability and security. In this highly competitive market. The next-generation network has a

**Corresponding author*: saranecedgl@gmail.com

A. Suresh, J. Ramkumar, M. Baskar and Ali Kashif Bashir (eds.) Resource Management in Advanced Wireless Networks, (19–44) © 2023 Scrivener Publishing LLC

single network that can transfer data from the source to the destination for a phone call, internet session, or video conference. These apps also provide value-added features such as real-time and non-real-time transaction and information systems.

Keywords: Visualization, network topology, computer network, next generation networks, tree-network topology, hybrid topology

2.1 Introduction

The Internet links hundreds of thousands of computers, cameras, monitoring devices, and Internet Protocol telecommunication devices, including a variety of programs and services such as the Web, email, and content material delivery networks. Hosts on the Internet are connected by hundreds of Internet service providers. Computer network management has been a big concern due to the exponential development and complexities of networks. It is critical to gain a graphical description of the network topology in order to track the network at the same time to react quickly to network errors. These systems can usually discover networked equipment such as routers, switches, servers, and so on in complex networks. Once the topology discovery is complete, these systems will display the topology discovery results in a graph. Typically, the elements are analogous to those shown in the topology. The Internet, in particular as a typical representative of computer networks, is a type of global open integration in which anybody can use several network applications in a heterogeneous network. Internet marketing, the advent of emerging network technology, and the exponential growth of Internet software have shown several flaws in the existing Internet system. The configurable network's visibility is steadily fading. Other problems, such as instability, hardware and software sophistication, and unpredictable wars, are also apparent and are now worsening. There is evidence that data networks have never seen growth and stability at this level of the Internet, and the structure in question has also shown unparalleled fragility and instability. Data network, the pattern in computer network creation in various directions, an outline of computer network complexity and the significance of researching the significance of developing next-generation network development criteria in the next-generation network structure. Increasing technological development and the provision of next-generation networks are crucial for public network operators, because public network operators can survive and develop by providing various value-added services at a lower cost, thus becoming an effective innovation Way to make a difference. In this highly competitive market. The next-generation network has a single network that can transfer data from the source to the destination for a phone call, internet session, or video conference. These apps also provide value-added features such as real-time and non-real-time transaction and information systems.

2.2 Topology Characterisitics

The Open Systems Interconnection Model (OSI model) is the most well-known network framework today, and it is the product of the International Organization for Standardization's attempts to implement open systems. This is a way of breaking down the coordination device into seven smaller components and these types of construction are called as layers. Each layer has functions that are similar in the sense that they can provide services to the upper layer while still accepting services from the lower layer [1]. At each level, the instance provides the highest level of service and requests the lowest level of service. The physical network topology emphasizes the system's hardware, including the cables that connect workstations, remote terminals, servers, and properties. The spatial topology of a system determines its physical relationship to other systems. This refers to the device's location [2]. The computer's devices are linked together by data cables. There are eight fundamental topologies. A direct association between two devices is referred to as a point-to-point topology (nodes). The value of a permanent point-to-point network is derived from the seamless connectivity of two endpoints. As an example, consider a machine linked to a printer. A more typical example is a host terminal connecting to the host's external interface.

When choosing which kind of topology to use to sustain business operations, there are several considerations to consider. This can include the number of employees, the office's physical size, and compatibility. Or simple topology groups, such as bus, star, ring, and mesh topologies, each with its own set of advantages. When calculating network size, take into account the number of devices that will be connected to the network. If only a few devices are needed, the bus topology can be sufficient. The following consideration is traffic control and network traffic. To server to use a large number of connected devices to handle a large amount of network traffic, a star topology may be a good adaptive approach. A central hub with Ethernet and unshielded twisted-pair cables is used to link point-to-point networks [1, 2]. There may be a full network that forms a high-speed backbone network or a partial network that connects computers to facilitate simultaneous connectivity. You can use a ring network, a closed-loop network, or a ring network to increase the security of the network topology. In this section, knowledge flows in only one direction along the topology. This also ensures that personal knowledge can be restricted to only those computers. If you need to progressively extend your network, you can experiment with combining simple topology types to construct hybrid types [3]. The current reputation of the computer network, and the assessment of pc community structure, and the significance of studies to build the next technology of community development wishes of the brand-new technology society structure.

2.2.1 Network Structure

Software, firmware, chip-level architecture, circuitry, and electrical pulses are all part of network engineering. The whole network model is split into many stages to simplify network architecture. Each level is responsible for a distinct role and is independent of the others. However, nearly all network activities depend on all of these layers in total. The layers communicate with one another and rely solely on one another for input and output data.

2.2.2 Modeling Network Layers

In layered architecture of Network Model, one whole network process is divided into small tasks. Each small task is then assigned to a particular layer which works dedicatedly to process the task only. Figure 2.1 shows the network model structure. Every layer does only specific work. Networking behavior is divided into small roles in the Network Model's

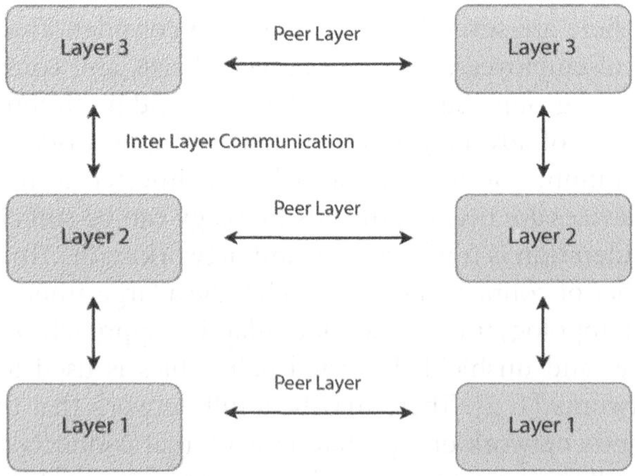

Figure 2.1 Network model structure.

layered structure. Each small assignment is then delegated to a specific layer that is committed to completing the project in the most efficient manner. Every layer does the most powerful, one-of-a-kind job. In a layered communication scheme, one layer of a group provides with the project being performed or to be executed by its peer layer on the same stage at a remote host [4]. The project is started either at the bottom layer or at the pinnacle optimum level.

If the task is started by the lowest sheet, the opposite direction is taken. Per layer collects all of the processes, protocols, and methods needed to complete the specific mission. The encapsulation header and tail are used by all layers to describe their counterparts.

2.2.3 Research Topology Challenges

The first difficult situation that this network layout will face is a financial and budget constraint. The cost of constructing the neighborhood gadget is limited and cannot be surpassed. This network architecture may be the solution for building networks within developing countries. This is due to the fact that this design can be made to be an integration group device, with true security and top of the line gadgets at low costs. The availability of system is needed to complete the installation of all group components [5]. Developing nations no longer have any of the equipment needed for a local area network, such as a router, backup system, and firewall device. The supply of gadgets is yet another challenge that must be dealt with when planning a society in developing nations.

Network systems that have been updated with new requirements and possible errors can occur on a regular basis. Despite being trained, network management users (particularly those in developing countries) are unaware of this. Many users of these countries/regions are unable to use the system's full capabilities. Network users in developed nations, for example, lack the network expertise to comprehend the idea of linking machines to a network so that they can use the same printer. Despite improvements in efficiency, dedicated printers for each machine are becoming more popular. This is a nebulous term, similar to sharing files and data among connected computers. Users would be more familiar with the network and its technologies than ever before if this network is properly built and used. Although network technicians in developing countries lack sufficient network competence and qualifications, their more experienced counterparts in developed countries can work hard to educate and transfer their information so that technical personnel in developed countries can gain greater autonomy and independence [4, 5]. Nowadays, network networks are typically static, resulting in unreliable data, higher overhead, and delayed representation of network changes. Additionally, as the domain itself changes, there is the issue of aggregating data across several domains. There is no automated updating to inform the inventory system of parallel adjustments when the inventory request is made over the network [6]. The network state and the inventory device display are not consistent. The availability of domain administrators or their attributes can have an effect on the final portrayal of the end-to-end topology. An inaccurate network view will be created if the management host is not automatically notified of changes.

2.3 Current Network Topologies

The section gives brief description of physical topologies. And introduces the logical topologies and finally conclusions with different structures and their use cases. There are different kind of topology we are going to see the most use cases and most build structures are the bus network topology, ring, star, mesh and tree are the types.

2.3.1 Bus Topology

A single cable connects each workstation in the bus topology. The line bus topology is another name for the bus topology. To put it another way, the interface port links all nodes computers and servers through a cable. The central cable serves as the network's backbone, and each workstation is linked to other devices by this bus. Along with the wire, the bus topology incorporates the transmitted code. When a message arrives at a router (node), the node checks the destination address found within the message to see whether it matches its own. If the deal with no longer matches, the node does nothing else. The biggest disadvantage of this community topology is that if the node deals with suits found within the packet, the node strategies the message [6]. The message is sent through the cable and is visible to all devices connected to the cable. Problems of security as information is disseminated across the community and a single point of vulnerability - whether the hub or spine fails.

2.3.2 Ring Topology

A ring topology connects all nodes or modules in a ring fashion. Data is sent from one device to the next, then through the entire ring circuit until it reaches the destination node. The only disadvantage is that if only one device is switched off, the network is still turned off. In this network, every tool is attached with the two devices on both aspect of it. There are devoted factor to factor hyperlinks a tool has with the gadgets at the both aspect of it. This shape paperwork a loop accordingly its miles referred to as ring topology. If a tool desires to ship facts to some other tool then it sends the facts in a single direction [6], every tool in ring topology has a repeater, if the acquired facts is meant for different tool then repeater forwards this facts till the meant tool gets it.

2.3.3 Star Topology

The star bus topology represents the hub propagation model in a computer network. Each host in a star bus network is connected to a central environment known as a node and also serves as a gateway. The network manager is also in charge of managing all network links as well as data flow between nodes connecting to the network. This is one of the more common approaches for connecting multiple parts [7]. The transmission lines that link the server, terminal nodes, and hubs form a star topology diagram. Before reaching its destination, information about the star bus topology can pass through the hub. Network and track all data coming in and going out of the network. It functions similarly to an information relay routed across a network. One of the most commonplace community configurations is a star topology, in which each device or node on a network links to a central hub. The significant disadvantage to this community topology is that if the primary node crashes, all computer systems connecting to that hub can be disconnected. One significant disadvantage of this group topology is the machine's reliance on the operation of the primary hub [6, 7]. When the primary center fails, the society becomes inoperable. The central hub and workstation will be shut down if a cable in the star bus topology fails. Just the damaged workstations are isolated, the remainder of the network remains operational. No, they should be attached to the impacted workstations so they are simple to handle, comprehend, and mount, as well as being inexpensive. The method is fast since they are made with customizable connectors and versatile cables. Complex routing protocols, such as network flexibility and service, can be easily managed.

2.3.4 Mesh Topology

A mesh topology is a network in which all nodes connect with one another in order to transmit data. This topology was created 30 years ago for military applications. They are, however, now commonly used in home automation, smart HVAC power, and smart buildings. The radio in many mesh networks will arbitrarily organize itself as required by using so-called special distance vectors. It is a reactive protocol that also has some active routing capability. This ensures that developers build certain routes and identify nodes that serve various functions, such as B. terminal nodes or repeaters. Endpoints, in their most basic nature, serve not only as goals and data bases, but also as relay points. Mesh networks operate in two ways: relaying data and flooding data. When you forward a message on a mesh network, the message follows a fixed direction, hopping from one node to the next before it arrives at its destination. Install these routes and ensure that they are available. The network must be continuously interconnected and optimized. This necessitates an active quest for damaged routes as well as the creation of self-healing algorithms for constructing routing tables [8]. Two layers of (MAC) traffic pass across the network to establish this path; mesh networks are less powerful than star networks. Flooding techniques are used in simpler mesh networks, in which data flows constantly across the network. When the module sees data using its own, it simply collects it. This is thanks to the time-to-live (TTL) value, which requires messages to travel a set number of hops before being removed. Bottlenecks and connectivity issues can arise if there are no nodes in the building. For Example, Zig Bee can only transfer data from one node to another over a distance of 30 to 50 feet. This might be a challenge if you don't have very thick sensor positioning in the house. A mesh network's spectrum is extremely limited. With the HVAC installer, who confessed to installing additional thermostats in the home?

2.4 Trends of Modern Networks

There have been several developments in communication generation in recent years. This greatly outperforms the computational power of small, resource-constrained handheld cellular devices. Various generations of communique generation have progressed. This resulted in new experiments for the communication of large quantities of statistics in specific transmitting media, as well as the design of different communication protocols. Another line of research focuses on maintaining a secure and error-free communication between sender and recipient in the presence of an eavesdropper. Several experiments have been conducted within the design of appropriate overlay networks to meet the communication demand of a large quantity of multimedia streaming statistics.

2.4.1 Network Link and Collabration

There are two important networks in use nowadays, the Public Switched Telecommunications Network (PSTN) and the Internet. Each server is set up to provide the linking services that it is capable of. The public telephone network is built on line switches, which offer very high service efficiency to major corporations. Toll-free calls, dispatcher systems, and customer calls are examples of advanced voice services. This network supports data services such as fax and e-mail through a ring link. Fixed bandwidth and low latency. A network that has been designed and tuned to accommodate well-managed, mathematically modelled Poisson voice traffic [9]. PSTN connections are provided by switches, switching machines, and intelligent networks. The cellular network, which mostly delivers mobile voice services, is currently interconnected to the public switched telephone network, which provides wired and wireless customers with worldwide connectivity.

It is also estimated that over the next eight to ten years, the number of Internet-connected devices will exceed the number of people on the planet. When you know how many mobile gadgets will have IP addresses in the coming years, this estimate gets more difficult to explain. The opportunity for machine-human communication and data sharing [10]. Since data traffic is superior to voice traffic and higher QoS values can be achieved in packet networks, especially for voice and other real-time services, it is preferable to link several networks across a single packet-based core network. This integration would allow new multimedia services, increase operators' capacity to satisfy their customers' varied needs, and reduce network operational costs. The evolution of network convergence is shown in Figure 2.2.

2.4.2 Disruptive Technologies

To explain technical advancement, a learning curve that measures the characteristics of the system over time may be used. If technological change is substantial and sustained, it has the potential to have a devastating effect on the economies it produces and represents. Photonics and wireless technologies are assisting in driving the market for converged networks. These architectures provide a low-cost foundation for the network elements that will make up the next generation of converged networks. It is all about technology, new architectures and algorithms, and QoS. As a result, project-based networks are appealing because they allow these networks to be carrier-grade [10]. Previous technological developments culminated in the development of processors and storage devices that are compatible with today's networks. Processor speed, random access memory (RAM)

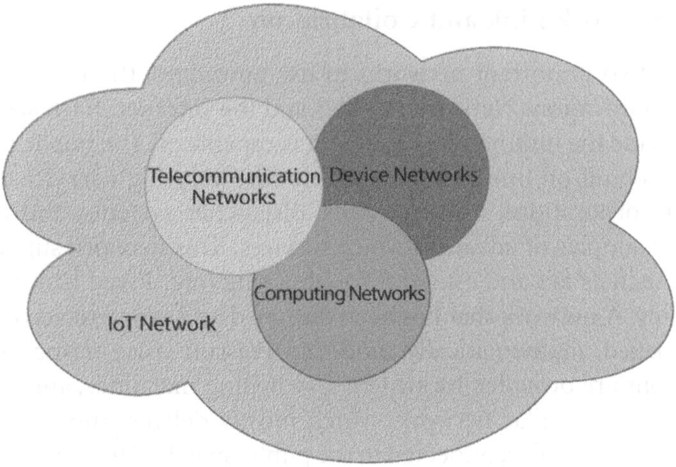

Figure 2.2 Evolution of network convergence.

capacity, and memory access speed, in particular, are all increasing at the same rate as optical fiber. The available speed of the end user. Complex management processes and memory are needed to serve fast data packets from high-speed data packets accessing network components. If the scalability and QoS concerns listed above are addressed, these developments would enable the new network's capability to surpass 100 to 250 times that of the old networks in the early 20's.

2.5 Architecture of Next Generation Networks

Next-generation networks use automated intelligent interfaces and multimedia data to link all, every computer, and every resource (regardless of distance, place, or time). Next-generation networks, as a medium, provide abundant networking and human-machine interactions that are more efficient. To make extensive use of resources that can bring meaning to human life in order to better life and develop new experiences. The website also serves as a "playground" for everyone to build and provide resources to others. The integrated intelligent interface can not only help non-commercial media (such as text, images, audio, video and other media) to convey information with various senses (including smell), but also eliminate language and technical barriers, such as written and oral expressions [11]. It has taste and touch, as well as images and sounds. The combination of a PC and the Internet as a forum now provides an astonishing range of information.

Communication, intellect, teaching, and entertainment are all conceivable, and contemporary technology is in use that can assess one's behavioral patterns and provide matching products and services, as well as communities of pastimes developing automatically, among other uses cases. This will be built upon by NGN. Aside from linking computers and distributing richer and context-aware information, it will network new devices and machines for the benefit of people, it intends to add value to people's lives by improving quality of life and enabling new lifestyle experiences.

Domain names for the provider stratum, distribution stratum, patron terminal interface capabilities, provider or utility data layer, control capabilities, and multiple networks include the NGN as shown in Figure 2.3. A consistent utility advent/execution environment [10]. It has clean provider introduction the use of utility for Application Programming Interfaces (API), which are referred to with high-level programming languages. It offers several sorts of vendors for provider or information, such as internal or third-party, trustworthy or untrusted suppliers. The difference between the service stratum and the distribution stratum, which allows for service expansion with little influence on the delivery layer.

2.5.1 Functionality of NGN Network

In the previous NGN architecture, there are more specifics in the administration section, which is essential to realize the customer-oriented process.

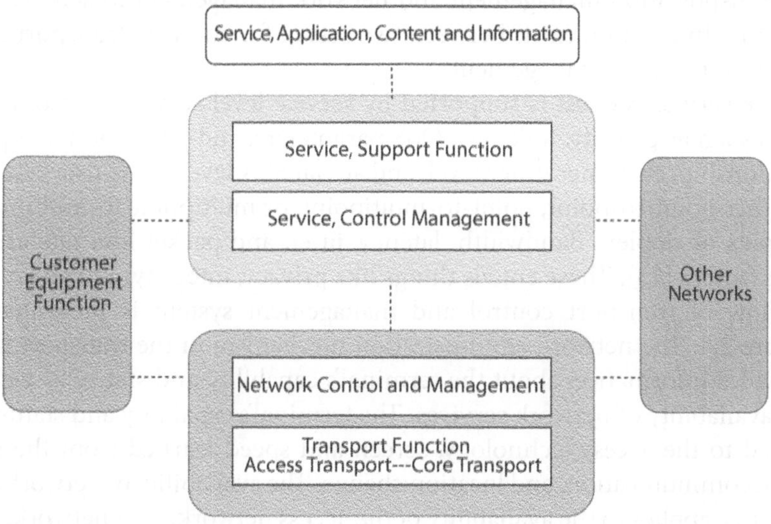

Figure 2.3 NGN architecture.

Furthermore, performance assessment is a passive process that happens only after a crisis has occurred. Real-time correspondence between the OSS and the BSS [12]. In theory, all consumers are handled the same. There are some primitive methods for prioritizing or separating resources, such as voice contact preceding data. However, in the network, these processes are almost not configurable, and there is no if-then operation. NGN requires network administration in order for OSS and client services to sustain the network in accordance with actual market plans, and the importance of allocating network capacity to specific users or services is higher, implying high-quality services. The corporate plan is the execution of the NGN operator's business strategy [12]. The management tasks of the ITU NGN useful design should be split into a network management part and a network service management part, and therefore the transmission management network management portion and the service management part should be linked.

2.5.2 Transport Control and Management

End-to-end links are identified by transmission management and control functions based on service specifications, terminal functionality and condition, as well as network and resource availability and will be wired to the network's terminal. The access network and terminal, as well as the access network and core network that will be linked, will provide quality and service security. Transportation management and control tasks are dependent on transportation management and network management to achieve this mission. Interaction exists between two sets of function are transport control and transport management.

The service request is supported by service level service control functions such as priority, link sort, QoS parameters, and protection. Supplier relationship, customer order level, and actual SLA level of the user relation is either point-to-point, point-to-multipoint, or multipoint-to-multipoint, simplex or duplex. Bandwidth, latency, jitter, and packet loss rate are all QoS factors [13]. These entails things like privacy, integrity, and so on.

Flow of transport control and management system is presented in Figure 2.4. The network administration mechanism in the transport layer provides information about the terminal's capability and status, as well as the availability of network services. The terminal's capability and status are linked to the access technology, screen and speed learned from the network communication, and location change. The availability of network and resource applies to the availability of the access network, core network, and

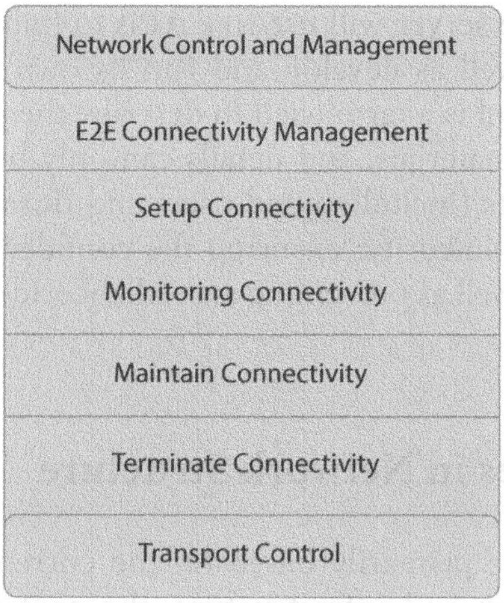

Figure 2.4 Flow of transport control and management.

network load in the end-to-end chain. This resource, as we all know, can be used to constantly track the access network and unified network.

2.5.3 Service Control Management

The service layer is in charge of developing and providing facilities, software, content, and data. It is split into two parts: service control and administration and service support. Create and maintain programs, software, content, and information for internal or external providers. Management of authentication protocols, activation procedures, and third-party services are all part of service support [12–14]. Activators can be added, saved, and deleted as desired. The key thing is that you will call the activator using an open API, where transparency means standardization.

The service management and control function, which is made up of service control and service control, is in charge of delivering NGN service. To create point-to-point requests, the service control allocates network services based on the priority of the requested service. The terminal is linked to guarantee that the service is supplied to the end user in accordance with business policies and that the client is fairly compensated for utilizing the service. These responsibilities include market strategy management, customer management, customer management systems, end-to-end QoS management, customer equipment governance, and compliance management. This service level requires servers, which may be provided by the network or service operator or by external vendors [14]. The key argument is that the server will use free APIs to find and use usable support resources, as well as develop and run its own service applications. Service access control is a term used to describe the process of restricting Services, software, materials, and details can only be accessed by signed and registered clients (including end users and third-party applications). Confidentiality and integrity: ensuring the confidentiality of stored and distributed data, as well as preventing surveillance, identity fraud, and data updates.

2.6 Challenges in Network Structure

While NGN may be primarily based on the current IP-based telecommunications network and infrastructure, the control and management architecture will be radically different, with a strong distinction between QoS-enabled networks/transmission domains and individuals. A service or domain programmer that operates in a distributed processing system. The pressures of deregulation, innovation, and accelerated technological growth, as well as the current NGN vision, would pose significant challenges to network and service activity, management, and maintenance.

2.6.1 Implementing Dynamic Topologies

It is anticipated that in NGN, equipment (particularly high-performance routers and switches) will become more programmable, and that more control software will be able to run directly on the equipment. SPP is subject to adjustment at any moment. Furthermore, communication between various network domains or service providers is dramatically enhanced. The complex design and topological framework of NGN will put standard configuration management approaches to the test, which are typically inefficient and necessitate excessive humanitarian effort [15]. NGN needs responsiveness and network-wide configurability to manage the rapidly evolving network topology (which can include thousands of distributed nodes).

2.6.2 Management Questions

A management team is needed to control the network. This teams (technical personnel) must have the necessary networking expertise and experience. When technicians lack network expertise, they introduce more bugs into the network infrastructure than they solve. Expensive networking infrastructure whether professionals don't know how to fix the network or don't have enough network expertise, they may save a lot of money by mending network equipment since the cost of the service is less than the cost of destroying other equipment if the system isn't repaired. Many network networks may have a minor issue that needs to be addressed, but novice technicians may exacerbate the problem without resolving it [15, 16]. The repair time is short, but incompetent technicians will halt the network fix for an extended period of time and might not be able to repair the problem at all, wasting a lot of valuable time for no reason. Experienced technicians can keep their information base up to date.

2.6.3 Consideration of Traffic Control

NGN is a packet-based network that is in charge of transmitting various networks on a single IP-based communication network, ranging from conventional voice calls to data, video, and multimedia applications. Optimized with differentiated offerings in order to improve and ensure service efficiency and reliability. To track and coordinate all relevant services, traffic management functions for separating services and NGN traffic must be provided. Since all network links have a set speed, traditional TMN does not directly address the issue of traffic management [16]. In NGN, comprehensive traffic pattern management and tracking has been an essential feature of NGN network operators and service providers.

2.6.4 Cost Management System

The budget is a critical consideration in the creation of local networks. This would have an effect on the health and safety of the network's equipment. Although the costly equipment is of high quality, it might not be available due to financial constraints. The network system's configuration has shifted from first-class equipment to very low-quality equipment. If there is sufficient money to design the network, higher-quality equipment would be used. The price determines the equipment's output. Fiber optic cables, for example, are the fastest cables used on the Internet. However, owing to the high cost of fiber optic cables, they are not used in the majority of network ventures [17]. Budget influences not only the efficiency of network equipment, but also their security. If they are permitted to use the increased budget, install new applications, such as strong antivirus software, on network computers. This improves the network's security. Numerous recovery systems and backup firewall devices may be used on the network to increase reliability if the network architecture has a sufficient budget. The network has been established between the node and the network.

2.6.5 Web Service Management

Web platforms provide promising support for a variety of remote services while transparently hiding technical deployment information, lowering integration costs. Web resources are made up of different building blocks that are built on XML and SOAP. The last one is a message with no state. SOAP messages function best for heterogeneous networks since they can be used with common web protocols like XML, HTTP, and TCP/IP connections. Web servers are an excellent option for systems due to the widespread use of HTTP and the ease of use of XML-based SOAP [16]. XML-based control methods have been around for a while, and numerous vendors are researching them. While the conventional government community does not support this quality, it can still be seen in certain vendors' goods.

2.7 Network Analysis and Routing Protocols

Network research, engineering, and construction have traditionally been viewed as an art that incorporates basic principles for analyzing and choosing network technologies. Understanding on how to successfully integrate technology, services, and protocols, as well as familiarity of what works and what does not, as well as network architecture selection However, as in most fields of art, the success of web design is normally determined by who is working, and the results are not repeatable. This may have been appropriate in the early days of networking, when networking was more of a sport than a critical resource, because it did not produce direct revenue. Both at home and overseas. They are regarded as the "key" to business success1 because they provide near-real-time access to global knowledge.

The Internet of Things is commonly classified into three levels: Level I comprise a series of interconnected instruments that monitor objects and their surroundings. The gateway node that collects data from the embedded system is represented by Level II. Smart phones are typically known as gateway nodes, which use more computing power than embedded units. Servers or data center's that store data obtained from gateway nodes for processing are examples of Level III sensors. The server or data center conducts dynamic analysis by creating a model based on the data obtained from the gateway node. Routing protocols are classified into several groups, each with its own set of operating criteria.

2.7.1 Destination Sequenced Distance Vector Protocol

The protocol is working. DSDV is a routing protocol that is built on sophisticated routing algorithms. Each wireless node must maintain and update the routing table, which includes the destination address, next hop, hop count, sequence number, and the first link, by using this routing mechanism. Each entry in the routing table includes a sequence number, which can be used to decide if those routes are old in order to prevent loopbacks. To ensure the full routing of all nodes, each node sends its routing table to its neighbors on a regular basis. If the network topology changes significantly, the routing table can experience significant changes [18]. The node will also constantly submit the updated routing table to its neighbors, ensuring that the routing table change has time and event-related characteristics.

2.7.2 Dynamic Source Routing

DSR is a reactive routing protocol as well. Every wireless node has a route buffer thanks to this routing mechanism. Each data packet's path information is directly stored in the header. A discovery mechanism is also used for the DSR protocol. Path to search routes on the fly. When you need a direction. The DSR route discovery procedure is close to that of the AODV protocol. When a node needs to send a information packet to another node, it first consults its routing table. If no routing entries to the destination host are discovered, the source host sends a routing message. (RREQ) is broadcast to any node. To create the path, the node that receives the RREQ responds to the route answer (RREP) on the road sign or the originating source. In contrast to the AODV protocol, when a routing entry moves through a specific hop, the hop ID is recorded in the RREQ routing protocol [19]. As a result, when the path to the terminal node is registered, information about the route given will be used.

2.7.3 AdHoc on Demand Distance Vector Routing

The AODV protocol is a routing protocol that is used to handle routing tables. Routing table information must be kept even for short-lived routes. Save the return to the host that sent the RREQ temporarily. The AODV protocol is a reactive routing protocol. When a wireless node needs to deliver a data packet to a specific node, it consults its routing table. If no path entries meet the target host, the host sends an RREQ (route request) packet in order to locate a new route [19]. The node that receives the RREQ first scans to see if the data packet's destination address matches its own. If not, it can see if the intermediate point has a path to the destination node. Modify according to the packaging's instructions, and then switch [18]. The data's are received from the relay points the RREQ information the destination address are stored in the routing table.

2.8 Evolution of Networks and Services Towards NGN

Different types of physical transmission networks will now operate in parallel, each with its own service and an architecture suitable for similar services. The fixed telecommunications network is primarily used to provide voice services to landline phones using twisted-pair cables. The network was initially designed to support voice services and increase mobility. The cable network was designed to use coaxial cables to transmit television and radio broadcasting services to televisions and radio receivers. The Internet's aim is to offer basic Internet access on the World Wide Web. Surfing, twitter, FTP, and remote connections to fixed telephone networks are all available.

This network and service ecosystem has become fragmented and is increasingly evolving as a result of technological changes and the economic market for consumers. Even if the changes vary from one network to the next and from one service provider to the next, the pattern of changes is discernible. Aside from native networks, each form of legacy network attempts to deliver simple Internet facilities such as web browsing, email, FTP, and remote login, as well as specialized Internet applications such as IPTV [19]. Each form of service provider is trying to maximize its ability to provide services to others, such as telecommunications networks that provide broadcast services, Internet service providers that provide telecommunications services, and wired networks that provide network services.

Many of these activities have marked the beginning of the community's transition to NGN. The NGN framework will bring together the fractured state of affairs of community and carrier, to rich offerings, software, contents, and records for end-users, to make the most of the existing community resources, and to launch new get entry to and middle community technologies. Access to services, apps, information, and data regardless of network. The serving community should be chosen entirely on the basis of the user subscription, carrier requirement, terminal capacity, and community usable resource availability [20]. There are a few conventional advantages, such as enabling feasible quit-to-quit compatibility when meeting QoS and security criteria. Providing distribution and handling services through a commonplace network that is independent of the underlying maritime community. However, there is still a long way to go to accomplish this NGN harmonization given the current state of things between culture and transportation. In general, all sorts of active participants nowadays face the prospect of becoming an NGN operator or a supplier. Individuals that adhere to the correct evolution approach are likely to be successful.

2.8.1 Significant Stages in Network and Services

The NGN are emerging from today's networks and services. The revolutionary steps in the services are all listed as network convergence and development. IP base network management system. Network integration and service extension.

2.8.2 Network Convergence and Development

Convergence in the service business offers clients with the same services from diverse networks. Customers may view the same TV shows on these networks via PSTN networks on PCs, cellular networks on mobile phones, or cable networks on television. This convergence of services in current networks signifies the capacity to deliver external services connected to different types of networks. Figure 2.5 shows the network convergence example with mobile and other networks. The technology used to deliver services may differ. QoS may differ as well. Despite the fact that each service has its own network Quality of Service needs, such as throughput, latency, jitter, data error rate, and so on. It is important to upgrade the existing network capability in order to collaborate with this service [21]. This is commonly referred to as a bottleneck. Higher access throughput and lower latency enable portability from low-speed to high-speed.

2.8.3 IP Base Network Management System

As a result of the shift to IP-based services, traditional proprietary network services will be replaced by IP-based services (such as telephone VoIP). This transformation decouples the service from the fundamental carriage network, which is essential for service availability regardless of access. Through a proper adaptor, the laptop may view the same TV show through landline, cellphone, and cable connections. Furthermore, this separation enables you to add or delete services without impacting the underlying

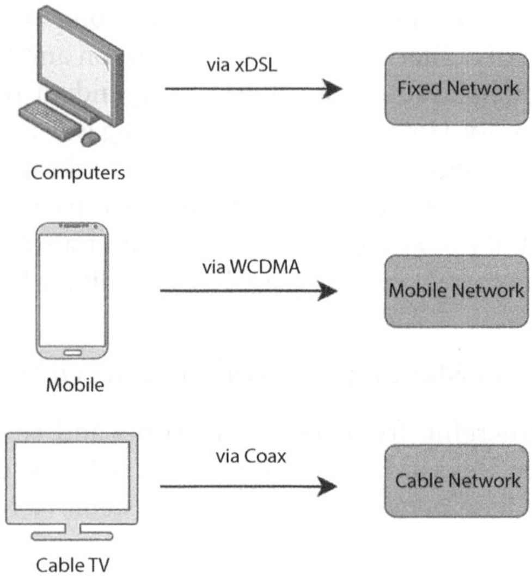

Figure 2.5 Network convergence example with mobile and other networks.

network, and changes to the underlying network have no effect on the service [21]. The converted service must be enhanced with sophisticated features. Traditional voice services used in telecommunications will be replaced by Multimedia over IP, which will be accessible across landline, cellphone, and cable systems. Status and availability information may be added to click-to-call, group calls, and the address book. Add/remove media based on the terminal's demands and capacity [22]. Traditional cable TV services are being converted to IPTV, which will be obtainable on fixed, cellular and cable setups.

Because IP-based services may have very diverse security and QoS needs, integrated QoS and prior security measures may not be enough; new methods must be identified and implemented to change the designated network into one of several competent core networks capable of providing QoS. The level of security is decided by the service needs, resulting in the managed IP network.

2.8.4 Service Extension and Network Integration

Network integration gives a single transmission control layer to all access networks and significant traffic networks, forming the NGN architecture's transmission layer. The transmission control plane is in charge of configuring and maintaining end-to-end connections to satisfy service needs by utilizing all forms of access networks and key transmission networks.

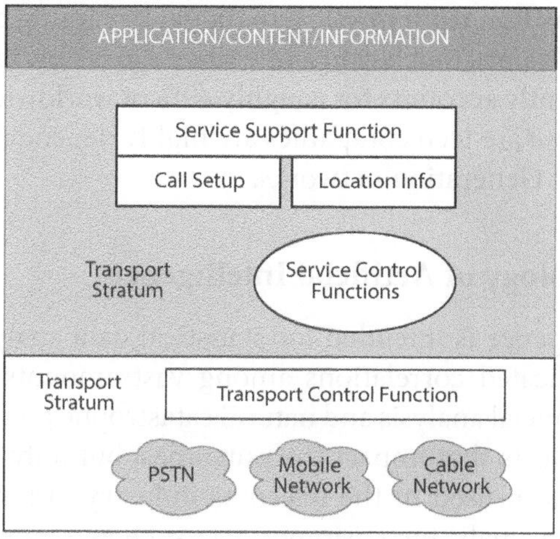

Figure 2.6 Next gen network service and transport stratum.

The conventional network is transformed into a clean channel of endways or partial connections. Endwise IP packets with the capacity to manage circulation contrarily based on QoS necessities, and in this common transmission control layer, traditional cables from landline, mobile networks are combined into a single transmission group known as end-to-end transmission [22, 23]. Terminate the connection while maintaining the needed QoS and security. Management functions, which are likewise placed on the transmission control plane, facilitate endpoint connections.

Figure 2.6 shows the next generation network service and transport stratum service extension entails hosting the service layer and open interfaces in order to design and offer effective services to end customers. A service layer can be formed by combining comparable functions of fixed networks, cellular networks, and wired networks.

2.9 Advanced Network and Service Management Technologies

As technology changes the content and volume of some goods and services, it also changes the content and volume of trade flows, some of which may have unanticipated consequences for trade flows. Renewable energy sources like solar and wind have lower demand than carbon fuels like coal and liquefied natural gas. As organizations such as video streaming businesses and music streaming providers establish streaming and subscription models, the transition from physical to digital streaming of personal movies, albums, and games that began a few years ago is now taking shape [23]. Streaming currently accounts for roughly 40% of worldwide music recording income. These are tech companies are highly dependable on the future network or Next Generation networks.

2.9.1 Technology of Artificial Intelligence

Artificial intelligence is intended for statistical data analysis, such as discovering complicated correlations among vast amounts of multidimensional data. Financial analysis and natural catastrophe prediction both have hundreds of aspects that impact the outcomes, but only statistically. This technique has been used in the telecoms industry to forecast churn rate by proportionally analyzing customer data such as gender, maturity level, employment, wealth, location of residence, native language and membership history, node usage, and service quality. Technology has recently been utilized to develop WCDMA wireless networks with self-adjustment and self-adjustment abilities. NGN claims to optimize network resources while ensuring a positive user experience [22]. AI technology is a potential technology that can help radio networks self-configure and self-optimize in a multi-radio environment, as well as improve active network performance.

These are two examples of creation approaches for WCDMA networks: self-optimizing and self-optimizing artificial intelligence. Each base station on the WCDMA network must be set with hundreds of parameters. These variables are influenced by topology, design, radio environment, traffic demand structure, network load, and service type. To achieve the optimum WCDMA network performance, these settings must be properly adjusted [23]. The following facts contribute to the complexity of WCDMA networks: coverage, capacity, and quality are all complicated and interconnected, and there is no clear link between meter settings and performance of network metrics.

The most difficult component is the connection between service consumption and network performance, which needs continuous WCDMA network design improvement [24]. Furthermore, with the concept of home/office base stations, the number of base stations in the radio network will increase in the future.

2.9.2 Smart Agent Technology

Intelligent agent technology is a customized technology capable of creating a virtual representation of each thing and learning/creating a profile depending on the item's behavior and actions. In the payment sector, for example, each cardholder, merchant, or individual is allocated an intelligent agent. Intelligent agents attached to items (such as cards or merchants) learn in real time from each completed transaction and develop their own distinct behavior over time. Smart agents aren't based on pre-programmed rules, and they don't aim to anticipate every conceivable scenario. Instead, they develop unique profiles for each creature and behave based on the objectives, observations, and information they get through interacting with others. To create a reliable virtual profile, each smart agent can extract all relevant data from multiple channels, regardless of type, format, or data source. Each profile is automatically and in real time updated, and the resulting data is transmitted via an intelligent agent. Smart agents may represent any organization and deliver world-class results while requiring minimum capital and operating resources [24]. The intelligent agent will check the data integrity automatically and execute functional training, data enrichment, and configuration files in the correct sequence. Intelligent agents, by focusing on updating configuration files depending on the actions and actions of entities, only keep relevant information and analytical data rather than the original input data they have processed, which may considerably compress memory.

2.10 Conclusion

The transition to ubiquitous cellular networking and networks, as well as some of the innovations that will serve as the foundation for the transition to next-generation wireless technologies. This includes smart radios and radio applications, as well as complex and flexible structures and technologies that provide modern paradigms for frequency allocation and management. Innovations of technology in Adaptive network are technologies like modulation, encoding, multiple access, media access, networking, and efficient adaptive networking allow next-generation wireless networks with incredibly high data speeds and efficient QoS. Technology and legislation are leading in a new age of network growth, paving the way for a converged next-generation network. It offers a variety of wired and wireless internet access mechanisms to all forms of customers. Fast access removes the bottleneck in advanced service access. Access modern and conventional ads for innovative network technologies made possible by interoperability of public switched telecommunications networks and new open application systems [25]. Wi-Fi technology software for the Internet of Things is generating a lot of interest in a variety of research fields. In reality, there is a lot of Wi-Fi technology available nowadays all over the world. It goes without saying that everybody has benefits, but no one is perfect. The question of which age is the best for most services remains unanswered. In reality, each software situation has unique requirements in terms of range, energy usage, throughput, and network topology. Furthermore, problems such as cost, ease of integration, and security must be considered.

References

1. Q. Liu and Q. Liu, "A Study on Topology in Computer Network," 2014 7th International Conference on Intelligent Computation Technology and Automation, 2014, pp. 45-48, doi: 10.1109/ICICTA.2014.18.
2. A Study And Analysis on Computer Network Topology For Data Communication Santanu Santra, Pinaki Pratim Acharjya,International Journal of Emerging Technology and Advanced Engineering, 2013.
3. Lim, Francis. (2016). A Review-Analysis of Network Topologies for Microenterprises. 175-180. 10.14257/astl.2016.135.42.
4. Yan Wang, Mengqin Xia, Fasheng Yi and Jiazhi Zeng, "Research on New Types of Network Architecture," Sixth International Conference on Parallel and Distributed Computing Applications and Technologies (PDCAT'05), 2005, pp. 149-153, doi: 10.1109/PDCAT.2005.205.
5. Challenges of Building End-to-end Network Topologies For Multi-service Networks Anne-Marie Bosneag and David Cleary, WINSYS 2007 - International Conference on Wireless Information Networks and Systems.
6. Geon Yoon, Dae Hyun Kwan, Soon Chang Kwon, Yong Oon Park, Young Joon Lee "Ring Topology-based Redundancy Ethernet for Industrial Network" SICE-ICASE International Joint Conference, pp. 1404–1407, 18-21 Oct. 2006.
7. Banerjee, S., Jain, V., Shah, S., "Regular multichip logical topologies for light wave networks", Communications Surveys & Tutorials, IEEE, On page(s): 2 - 18 Volume: 2, Issue: 1, First Quarter 1.
8. D. C. Dowden, R. D. Gitlin and R. L. Martin, "Next-generation networks," in Bell Labs Technical Journal, vol. 3, no. 4, pp. 3-14, Oct.-Dec. 1998, doi: 10.1002/bltj.2125.
9. Berezdivin, Robert & Breinig, Robert & Topp, Randy. (2002). Next-generation wireless communications concepts and technologies. Communications Magazine, IEEE. 40. 108-116. 10.1109/35.989768.
10. M. T. Alam, Z. D. Wu, "Performance Analysis of SIP-Based Session Establishments Over IMS." The IASTED International Conference on Wireless Networks and Emerging Technologies, Banff, Alberta, Canada, pp: 178-183, Jul 2005.

11. Ernst Ulrich von Weizsäcker, Amory and Hunter Lovins. Factor Four – Doubling Wealth, Halving Resource Use. Earthscan: London, 1998.
12. Li, Mo & Sandrasegaran, Kumbesan. (2005). Network management challenges for next generation networks. Proceedings - Conference on Local Computer Networks, LCN. 2005. 6 pp. - 598. 10.1109/LCN.2005.93.
13. Zhu, Hua, *et al.* "A survey of quality of service in IEEE 802.11 networks." IEEE Wireless Communications 11.4 (2004): 6-14.
14. S. Sankaran and R. Sridhar, "Modeling and analysis of routing in IoT networks," 2015 International Conference on Computing and Network Communications (CoCoNet), 2015, pp. 649-655, doi: 10.1109/CoCoNet.2015.7411258.
15. Sriram Sankaran, "Modeling the performance of IoT networks", Advanced Networks and Telecommunications Systems (ANTS) 2016 IEEE International Conference on, pp. 1-6, 2016.
16. Jian Cao, Xingwei Wang, Min Huang, Xinhao Zhou, "A Mobility-Supported Routing Mechanism in Industrial IoT Networks", Access IEEE, vol. 7, pp. 25603-25615, 2019.
17. M. Saravanan, J. Ajayan, R. Maheswar, and P. Eswaran, An overview of architecture & applications of IoT based healthxare system, book: Smart healthcare systems: Applications, services, and challenges, Wiley Telecom, 2022.
18. Kalpana, V., Mishra, D.K., Chanthirasekaran, K., Haldorai, A., Nath, S.S., and Saraswat, B.K. On reducing energy cost consumption in heterogeneous cellular networks using optimal time constraint algorithm, Optik, 270, art. no. 170008, 2022.
19. C. A. Thekkath, T. D. Nguyen, E. Moy and E. D. Lazowska, "Implementing network protocols at user level," in IEEE/ACM Transactions on Networking, vol. 1, no. 5, pp. 554-565, Oct. 1993.
20. M. Yao *et al.*, "Artificial Intelligence Defined 5G Radio Access Networks", *IEEE Commun. Mag.*, vol. 57, no. 3, pp. 14-20, Mar. 2019.
21. M. Chen, W. Saad and C. Yin, "Liquid State Machine Learning for Resource and Cache Management in LTE-U Unmanned Aerial Vehicle (UAV) Networks", *IEEE Trans. Wireless Commun.*, vol. 18, no. 3, pp. 1504-17, Jan. 2019.
22. P. Francois, C. Filsfils, J. Evans and O. Bonaventure, Achieving Sub-Second IGP Convergence in Large IP Networks. ACM SIGCOMM Computer Communication Review, Vol.15, Issue 3, pp. 35-44, 2005.
23. W. Chien, C. Lai, M. S. Hossain and G. Muhammad, "Heterogeneous space and terrestrial integrated networks for iot: Architecture and challenges", *IEEE Network*, vol. 33, no. 1, pp. 15-21, 2019.
24. P. Wang, J. Zhang, X. Zhang, Z. Yan, B. G. Evans and W. Wang, "Convergence of satellite and terrestrial networks: A comprehensive survey", *IEEE Access*, vol. 8, pp. 5550-5588, 2020.
25. Pau G, Chaudet C, Zhao D, Collotta M. Next Generation Wireless Technologies for Internet of Things. Sensors (Basel). 2018;18(1):221, 2018.

3

Evolution of Next Generation Networks and Its Contribution Towards Industry 5.0

Kiruthiga Devi M.[1*] **and Padma Priya M.**[2†]

[1]*Department of IT, Dr. M.G.R. Educational and Research Institute, Chennai, India*

[2]*Department of Computer Science and Engineering, Sri Sairam Engineering College, Chennai, India*

Abstract

Next-Generation Network has a major impact on the telecommunication sector as it provides convergence of network and services. 5G networks are faster than the previous generation network and will meet the increasing demand of new consumers and business needs efficiently. NGN has become an emerging technology since it is a scalable, flexible, and more reliable network. In a very short span of time the wireless communication network has transited from 1G to 6G. As 5G networks have become more widely available, they will be a major driver of Artificial Intelligence (AI), Internet of Things (IoT), driverless vehicles, blockchain, augmented reality, and a potential breakthrough. 5G has reached unimaginable heights because of fast link, cost-effective, low latency, efficient data transmission, broader connectivity, and coverage. The aim of this article is to conduct a review of the literature and investigation of 5G requirements, network architecture, and its contributions towards its society. We also discuss the evolution and development of wireless technology through various generations, emphasizing the significance of 5G transformative networks, examining the challenges, and exploring the 5G applications in various sectors such as Healthcare, Entertainment, manufacturing industries, Automobile, Intelligent automation, and industry digitization. It also elaborates on shifting towards 6G and its system architecture. The sixth-generation (6G) wireless network will offer terrestrial, maritime networks, and airborne. Additionally, it will be extremely fast, and be able to take care of large number of devices with very low latency requirements. 6G networks would be much more heterogeneous than their predecessors, allowing for technologies such as virtual

**Corresponding author*: kiruthigaprofessor89@gmail.com

†*Corresponding author*: padmapriya.cse@sairam.edu.in

A. Suresh, J. Ramkumar, M. Baskar and Ali Kashif Bashir (eds.) Resource Management in Advanced Wireless Networks, (45–80) © 2023 Scrivener Publishing LLC

and augmented reality (VR/AR), ubiquitous instant messaging, widespread intelligence, and the internet of abilities that have yet to be seen in the mobile world. 7G should have no problem covering bandwidth or data capacity. Space roaming is at the core of the seventh generation of mobile wireless networks. It is becoming more and more necessary to reach an entirely wireless world where information can be accessed at any time and from any location, with lower costs, better quality, and access at any place. 7G will have no issues with data capacity coverage or handoff when it completes all of its phases. Users will be subject to only one demand during that period, namely, the cost of mobile phone calls and related services.

***Keywords*:** Next-generation network, Internet of Things, network architecture, intelligent automation, healthcare, virtual and augmented reality, ubiquitous instant messaging

3.1 Introduction

Next-Generation Network (NGN) is a Packet-based and high-speed Network where the data, voice messages, and other media are encapsulated and transferred as Packets. It's a layered architecture, provides Telecom services and unlimited access to cellular users through disparate service providers [44]. It utilizes multiple broadbands and supplies consistent and pervasive services to customers, providing point-to-point quality of services and transparency. It supports a large range of applications and services with the adoption of a multi-service network. The Public switched telephone network is integrated with the Public switched data network. Flexibility in NGN is achieved through the use of Soft-switches. Soft switches are used to connect PSTN and IP networks in order to provide efficient telephone connectivity [46]. NGN supports varieties of media such as text files, voice messages, videos etc. It provides access to multiple devices such as laptops, Personal Digital assistants, home appliances, and many other handheld devices. It connects WLAN, cellular, Bluetooth, wired network, and devices.

The Evolution of Cellular Wireless Networks

- **1G:** First-generation mobile communication systems (1G) were introduced in 1970's and released in 1979. It's an Analogue mobile phone system and uses AMPS (Advanced Mobile Phone System) technology. One of the major challenges in the 1G era was poor voice quality and reliability. It had limited capacity and provided no security.

- **2G:** 2G released in 1991 was the first wireless digital transmission. It was developed mainly for fax, MMS, voice transfer, and SMS services. It utilizes Computerized signals for voice communication. The Global Systems for Mobile Communication (GSM) enables the transfer of data on top of voice at 30-35 kbps. The introduction of GPRS as an enhanced data service carried the 2G into the 2.5G space. A GPRS connection can transfer data at a maximum speed of 170Kbps [14]. The data per rate 4 times increased from 170kbps to 500kbps. 2G systems had the major drawback of not being able to handle complex data and meeting user requirements.
- **3G:** Introduced in 1998 and released in 2002, 3G mobile framework involved UMTS (Universal Mobile Telecommunications System). Device speeds are estimated at around 2Mbps, and mobile phone speeds are estimated at about 384 kbps. 3G involving CDMA (code division multiple access) and Edge technology. It also introduced media streaming and achieved seamless Global roaming. 3G further extended to 3.5G by involving High-fast packet access which increased data transmission rate up to 14mbps. 3G networks demanded 3G compatible devices and the cost of upgrading to 3G devices and power consumption was too high.
- **4G:** 4G network released in 2009 offering LTE (Long Term Evolution) and improved downloading rate of up to 100Mbps.It provided a quality that enriched user experience with advanced gaming services, Video Conferencing, HD mobile television. 4G LTE (Long Term Evolution) system design resulted in the reduction of high-speed data transmission latency, which ultimately increased network efficiency and speeds. The major difficulties faced by a 4G, LTE network include complex hardware, a large number of radio transmitters, and low battery life [65].
- **5G:** 5G refers to the Fifth-Generation network, launched in the year 2019. It provides faster data rates, lower latency compared to previous generation networks, and higher connectivity. It allows us to connect more devices simultaneously providing a user with a real-time experience. It will lead to smaller IoT devices and can reduce the cost involved in manufacturing. 5G can enable new services that can transform low-latency links, industries with ultra-reliable [41].

	1G	2G	3G	4G	5G
RELEASED	1979	1991	2002	2009	2019
STANDARDS	NMT, AMPS & TACS	GSM & CDMA	UMTS & EV-DO	LTE	5G
CAPABILITIES	Analog Voice	• Digital Voice • Encrypted Communication • Limited Roaming • SMS & MMS	• Mobile Broadband • Locating Services • Multimedia • Streaming • Seamless Global Roaming	• High Speed Mobile Internet • IP-based packet switching • HD multimedia streaming • Seamless global roaming	• Private networks • IoT Ready • Massive machine type • Ultra low-latency • Ultra high reliability • Millimeter wave support
EXTENSIONS		GPRS (2.5G) CDMA 2000 (2.5G) EDGE (2.75G)	HSPA+(3.5G)	Feature extension through new category releases	Feature extension through new category releases
	.0024 Mbit/s	.064 Mbit/s	42 Mbit/s	10000 Mbit/s	10000 Mbit/s
	No impact on industrial application	• Remote control/ Tele control • Text messages from and to remote machines	• Video monitoring • Remote access to machines • Remote condition monitoring	• Mobile service technicians • Service via smart phones • Wireless background	• Autonomous logistics • Autonomous machines • Assisted work • Wireless back • Edge computing

Figure 3.1 Evolution of cellular wireless networks.

- **6G:** The sixth-generation network that is recently under construction, supports cellular data networks. It is significantly faster than 5G [2]. The network will be a broadband cellular service in which geographical areas are divided into smaller units known as cells. It will exhibit more heterogeneity than the previous generation networks. It will support Virtual/Augmented Reality, the Internet of skills, Ubiquitous computing, and pervasive intelligence applications [21]. The evolution of the cellular network is shown in Figure 3.1.

3.2 Networks of 4G

The first version of the Long-Term Evolution LTE standard was widely adopted for 4G networks, which is an IP-based network technology that does not allow circuit-switched networks. Among them are 4G networks, which combine fixed Internet connections with radio access networks. The ITU-R organization set the International Mobile Telecommunications

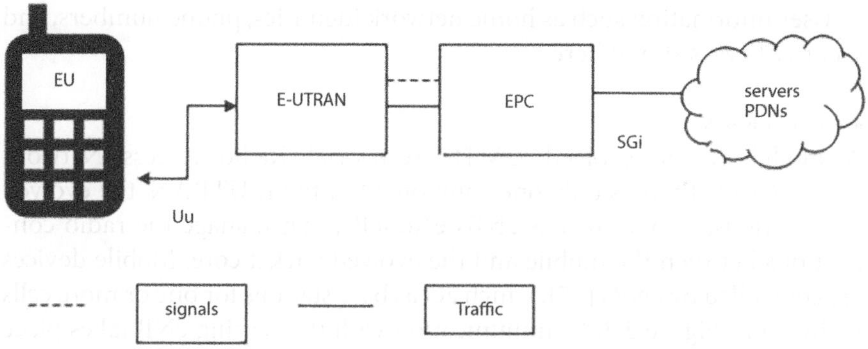

Figure 3.2 Interface of LTE.

Advanced IMT-Advanced specifications for 4G standards in 2009, requiring peak speeds of 100 Mbit per sec for high mobility communication (such as from trains and autos) and 1 Gbit per sec for low mobility communication. Among the technologies that are essential for 4G and beyond is Open Wireless Architectural OWA, which supports a wide range of wireless air interfaces. One of the technologies utilized in 4G is Software Defined Radio (SDR). Configuration or definition of the radio is done using SDR, which allows the radio to function in a number of different applications. In addition, base stations and terminals will be multiband and multi-stack [38]. In the future, the terminal will adapt to the available radio access technology with air interface; now, the infrastructure does this. The UMTS Terrestrial Radio Access Network (E-UTRAN) [65]. The User Equipment (UE), and the Evolved Packet Core (EPC) are the primary components of the LTE high-level network architecture is shown in Figure 3.2.

- **User Equipment**

LTE user equipment has the same internal architecture as GSM and UMTS equipment referred to as Mobile Equipment (ME). The mobile equipment is comprised of the following important modules:

- Mobile Termination (MT): All the communication functions are handled.
- Terminal Equipment (TE): Data streams at these terminals.
- Universal Integrated Circuit Card (UICC): LTE equipment SIM cards are also called LTE SIM cards. An application called Universal Subscriber Identity Module (USIM) runs on it.

User information such as home network identities, phone numbers, and security keys is stored here.

- E-UTRAN

A model of the evolved UMTS Terrestrial Radio Access Network (E-UTRAN). There is only one component of the E-UTRAN, the evolved base stations, also known as eNBs/eNodeBs, that manage the radio connections between the mobile and the evolved packet core. Mobile devices are controlled by eNBs [37], which act as base stations for one or more cells is shown in Figure 3.3. Communication with the serving eNB takes place at the base station.

One Base station and one cell at a time with LTE Mobile communication, eNB's primary functions

- In order for the eBN to transmit and receive the radio broadcast, it uses the LTE air interface's analog and digital signal processing capabilities.
- By sending signaling messages to its mobiles, such as handover directives, the eNB conducts all its operations at the lowest level.

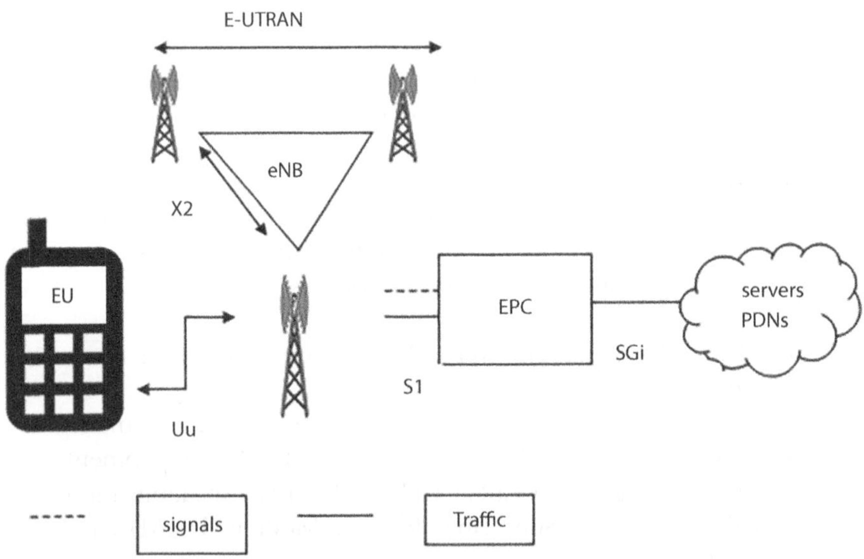

Figure 3.3 Architecture of E-UTRAN.

During the changeover, the S1 interface connects each eBN with the EPC, and the X2 interface connects it with neighboring base stations, which is primarily used for signaling and packet forwarding. An eNB (HeNB) purchased by the user is a femtocell base station that provides coverage within the home. eNBs belong to closed subscriber groups (CSGs) and are only accessible by mobile phones that have closed subscriber group SIMs.

- Evolved Packet Core

Below is a diagram of the Evolved Packet Core (EPC) architecture. To keep things simple, a few more components haven't been depicted in the diagram. The Earthquake and Tsunami Warning System (ETWS), the Equipment Identity Register (EIR), and the Policy Control and Charging Rules Function are some of these components (PCRF) is shown in Figure 3.4.

This architecture has a description of each component:

- Home Subscriber Servers (HSS) are centralized databases containing information about a network operator's subscribers. The protocol was carried over from GSM and UMTS.
- Using the SGi interface, the Packet Data Network (PDN) Gateway (P-GW) communicates with the outside world, i.e., packet data networks PDN. Each packet data network is identified by an access point name (APN). PDN gateways

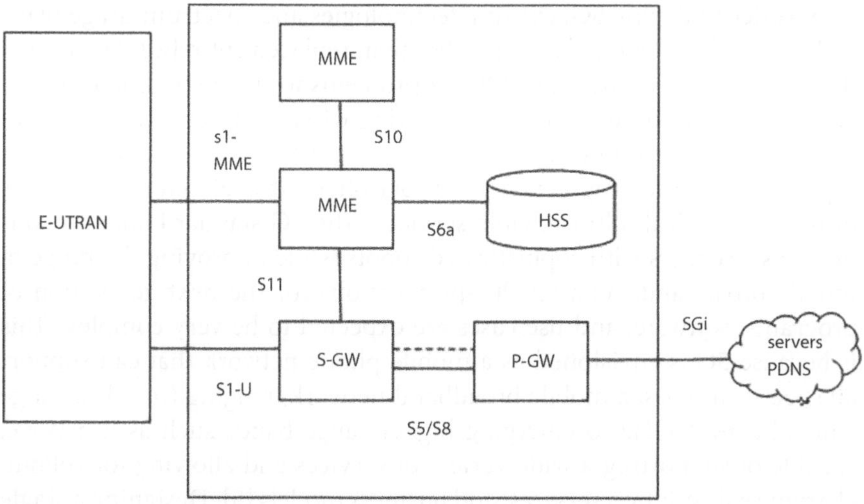

Figure 3.4 The architecture of evolved packet core.

support UMTS and GSM networks alongside GPRS support nodes (GGSN) and serving GPRS support nodes (SGSN).
- PDN gateways receive data from the base station through a serving gateway (S-GW).
- In the picture above, the Policy Control and Charging Rules Function (PCRF) is not visible, but it is responsible for directing the Policy Control Enforcement Function (PCEF), located in the Policy Governance Wireless.

A serving/PDN gateway interaction occurs via S5/S8. If the two devices are on the same network, S5 is used, and if they are on different networks, S8 is used.

3.3 5G and its Vision

The transition from 1G to 4G was marked by a change in channel access methods, such as FDMA, OFDMA, CDMA, TDMA, as well as enhanced modulation and coding schemes. However, unlike previous technologies, there is no guarantee that 5G will provide a single air interface depend on a single radio access technology (RAT). Some industry insiders envision 5G as a "network of networks," with a range of air interfaces, access node groups, protocols, frequency bands, and network forms. Current communication systems have reached their Shannon limit, and more new devices require wireless connectivity [17].

In order to deploy 5G, existing technologies and spectrum usage plans will need to be integrated better rather than replaced entirely. If 5G follows this forecast, one of the most difficult problems would be the smooth convergence of improved iterations of existing wireless systems with complementary new communication networks [18, 19]. Higher data speeds and capacity aren't the only benefits that 5G offers. It is aimed at new types of mission-critical, ultra-reliable services. The 5G standard aims to connect basic sensors with sophisticated robots while improving the range of mobile broadband. As a result, specifications for the next generation of programs, software, and use cases are expected to be very complex. This is because 5G is envisioned as a mobile phone network that can support all spectrum types, a mobile broadband network, ranging from low range bands below 1 GHz to emerging higher range bands such as mmWave, capable of supporting a wide variety of services and allowing for rollout, charging subscription services, and business models [5]. Designing a single air interface that is flexible and adaptable across all spectrums and service of types is key.

3.3.1 Requirements for 5G

According to the International Telecommunication Union (ITU), mobile broadband networks, large machine-style communications, and ultra-reliable and low latency communications are types of service scenarios that would be provided in 5G [54]. Due to the different types of devices and applications scenarios, highly complex networks need to not only maintain Increasing throughput, but must also maintain low latency in an energy consumption efficiency scheme, data transmission, and improved scalability to handle a vast number of devices [1].

- High data rates

While users can stream HD video content at maximum rates (8–15 Mbps), some applications, such as 4K ultra-HD streaming, need even higher data rates, 3D content, high-definition gaming about 25 Mbps, to provide users with a satisfactory experience. As a result of these new devices, 5G networks are projected to provide peak data rates of 10 Gbps, which is a 100-fold improvement over existing 4G networks. A wide range of technologies, including millimeter-wave communications, wireless software-defined networking and large MIMO networks, among others, will meet this demand for an increased data rate [23, 25].

- Low latency

There is potential to improve latency to a level of about 1 millisecond, to accommodate new emerging technologies such as tactile Internet, multi-player gaming, and virtual reality. In order to control virtual and real objects by interacting with them in a standard way without triggering cyber-sickness, latency is needed in such applications. The estimated latency required to make these applications work is about 1 millisecond.

- Low energy consumption

A 5G network is capable of connecting IoT devices and sensors to send information from a specific area to a centralized server. These are typically low-power, long-lasting, and low-cost devices. If these devices are not always connected to the base station and are only turned on occasionally, they cannot afford to have them synchronize all of the time since synchronization consumes much more energy than transmitting data. There will be more base stations to support smart devices as the number of connected devices increases. It is therefore essential that base stations boost their energy efficiency since even a modest increase in efficiency could lead to considerable energy savings on a large scale.

- High scalability

Mobile devices increasingly connect to next-generation wireless networks, so network scalability becomes an essential component. In order to effectively manage this increase in device count, a highly scalable network is needed. Scalability is also important for current and emerging applications like autonomous vehicles, IoT services, etc. It takes changes at every layer of the network, from the radio access layer to the core network, to ensure network scalability and virtualization.

- Improved connectivity and reliability

For a better user experience, coverage and handover efficiency should be improved, especially when using millimeter wave spectrum. As the density of base stations and the number of linked devices increases, the number of handovers the base station must manage will increase by at least two orders of magnitude. This demand requires the development of new handover algorithms and techniques that provide enhanced coverage at the edge of a cell. Another problem that is relevant to the handover is authentication and privacy issues. In 5G systems, the time it takes to connect to the authentication server is hundreds of milliseconds, making it inefficient. As a result, 5G faces a huge challenge in preserving connectivity. The criteria for high reliability and connectivity should always be guaranteed for mission-critical services.

- Improved security

Security requirements for 5G not only allow trustworthy users access, but enhance overall network security, addressing concerns about authentication, advanced encryption, authorization, and accounting. It was not possible for 4G networks to create consistent standards to protect users' personal information, but 5G networks will resolve this issue fully. As the Internet of Things (IoT) enters its prime in 5G networks, authentication, accounting, and authorization (AAA) processes. Providing fine-grained security mechanisms for interconnected devices is crucial. Network providers, standardization bodies, and equipment manufacturers can collaborate on the development of services, devices, and protocols that can protect subscribers of 5G wireless networks [3].

3.3.2 Network Architecture

The primary objective of previous generations of mobile networks was simply to provide their users with fast, reliable mobile data services. In addition to providing end-users with multiple connectivity channels and multi-layer networks, the spectrum of 5G offers an extensive range of wireless services [20]. An effective platform for 5G involves a variety of evolving technologies that are adaptable, seamless, and complex. The radio access networks (RANs) of 5G are no longer restricted by base station proximity or complex infrastructure [57]. As a result of 5G, RANs will be disaggregated, scalable, and virtualized, which will enable additional data access points. The 3rd Generation Partnership Project (3GPP) tackles telecommunications technologies such as radio access networks (RAN), service capabilities, and core transport networks. In an effort to achieve a much more service-oriented network architecture than the previous generation, the 3GPP has released the full set of 5G device requirements. In addition to modularity, reusability, and self-contained networks, other concepts that are required by 3GPP specifications for the design of 5G networks are modularity, self-contained , and reusability networks.

Its core architecture allows for improved throughput and 5G demands and is the basis for the latest 5G specification. In accordance with 3GPP specifications, the new 5G core employs a cloud-aligned, service-based architecture (SBA) that covers 5G functions and interactions, protection, session management, including authentication, and traffic aggregation from end devices. As one of the core design elements of 5G, Network Function Virtualization (NFV) is emphasized as an integral feature, with virtualized software functions implemented by way of the Multi-access Edge Computing (MEC) infrastructure [60]. The user plane direction consists of Radio Access Network (RAN), User Equipment (UE), Data Network (DN), User Plane Function (UPF) are not as evident as in the control plane, the control plane is modified in the way mobility and session management is managed in two separate entities, Session Management Function (SMF), and Core Access and Mobility Management Function (AMF). The service-based architecture provides the reference points between SMF, AMF, Authentication Server Function (AUSF), Policy Control Function (PCF), and Unified Data Management (UDM) shown in Figure 3.5.

Using this model, a network function can either offer service on-demand or subscribe to it. bile devise - The 5GC can now accept Non-Access Stratum (NAS) messages for Session and Mobility Management, which was previously not possible. As part of the 5G network, UEs are being able to communicate via network slices, which consist of control plane and user plane

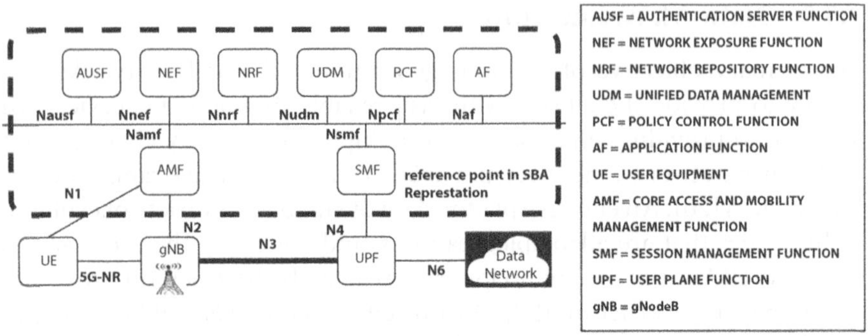

Figure 3.5 3GPP network architecture.

components with varied capabilities. A key feature of the 5G Core (5GC) is its ability to support non-3GPP [3] accesses such as Wi-Fi, i.e., end-user equipment (UE) - the protocol for NAS [12] related to mobility and session control messages. UEs on non-3GPP networks must be authenticated with an AMF over 3GPP networks using a new security procedure. In order for a UE to communicate with the network and AMF, finding ways to test the N1 and N2 interface is essential [27]. Solver supports both the N1 and N2 interfaces, enabling unique, bespoke testing at scale for 5G service assurance and node validation. New radio frequency ranges have been allocated to the 5G system (NR). Generally speaking, millimeter waves are within the range of frequencies from 30 GHz to 300 GHz and have wavelengths of 1-10 mm. 5G deployment is now taking place in various parts of the world at frequencies ranging between 24 GHz and 100 GHz. Additionally, to millimeter waves, underutilized UHF frequencies ranging from 300 MHz to 3 GHz are being repurposed for 5G [62, 63]. Having higher bandwidths but shorter ranges, higher frequencies can be adapted to particular applications. For long-distance communications, millimeter-wave frequencies are not suitable since they are inefficient in heavily populated areas. Within these 5G frequency bands for high and low-speed mobile, each carrier is carving out a piece of spectrum for their own unique use.

3.3.3 Economic Contributions and Networks Intelligent Automation of 5G

Digitalization has been adopted by manufacturers due to multiple reasons, including increasing sales resulting from better customer service, beating the competition, increasing demand, decreasing costs by increasing productivity and production, and reducing safety and security risks [6]. Devices with long battery life are of low cost, resilient, ultra-reliable, and instantaneous. Asset monitoring is important in today's ever-changing supply chains.

3.3.3.1 Internet of Things (IoT): Artificial Intelligent Plus 5Generation is a Smart IoT

"The Internet of Things" (IoT) assembles sensors and computers into a network that connects them to the Internet and to other networks [59]. The theory is based on the assumption that "all that can be linked will be connected. "For example, washing machines, wearable devices, coffee makers, cell phones, etc.

IoT architecture and working principle

As shown in Figure 3.6, the architecture is designed of eight interconnected layers that can share data in both directions [28]. The second and fifth layers are divided into two and three sub layers, and the protection layer guards them all [4].

Physical Device Layer

Wireless sensors, actuators, and controllers make up this layer. In all architectures, physical devices represent a common layer. Smaller components, such as Nano-Chips, will be used in this layer to boost computational processing capacity while lowering power consumption. A large amount of preprocessed data can be generated by nano-chips, which is ideal for Big Data used at the analytic layer.

Communication Layer

It comprises two sub-layers, Device-To-Device communication (D2D) and connectivity. Nodes Physical devices now have their own personality and identity, as well as the ability to produce their own data. By forming a HetNet, these devices can communicate with each other, which improves the efficiency and capabilities of IoT systems. This sub-layer makes use of the most up-to-date wireless sensor network (WSN) communication protocols. Clustering can be achieved by the nodes, and they can also choose a cluster head for proper networking. The proposed 5G-IoT architecture considers 5G-Plus-HetNet a viable technology solution for low-latency connectivity.

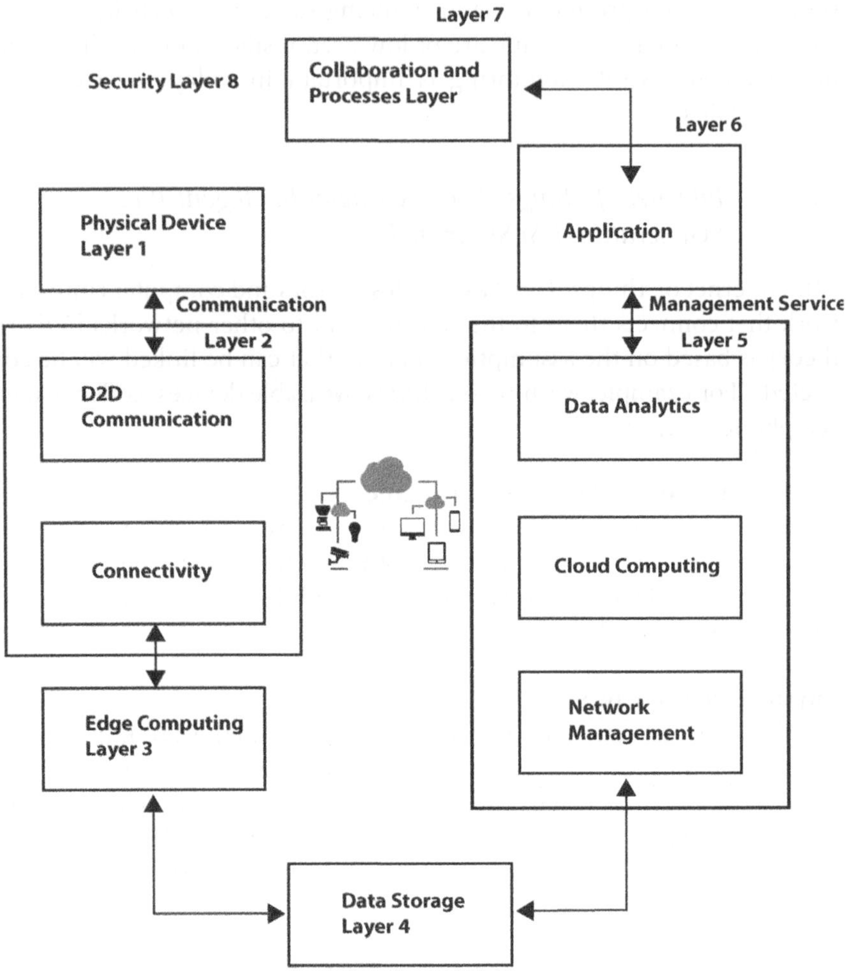

Figure 3.6 5G-IoT architecture.

Connectivity sub-layer

Devices are interconnected with the communication systems, such as the BS, in this sub-layer. They also use the Intranet link to the storage unit to send and review data via the centers. Due to significant communication latency, this IoT sub-layer is currently unable to support a large volume of data in real-time, and data sharing for a wide variety of data types is impractical for applications such as autonomous vehicles [29]. Implementation of 5G will significantly improve the efficiency, reliability, and agility of this sublayer in the near future. SSIM technology enables IoT devices to select appropriate spectrum with minimal interference. Aspects of SSIM can be used to take advantage of spectrum-sharing applications using cognitive radios. Communication robustness, Low latency, and support for a wide range of data types are all provided by this sub-layer built-in architecture.

Edge Computing Layer

In this layer, nodes process data in order to make decisions. 5G technology and the proliferation of mobile devices (like tablets, smart phones) means Multi-access edge computing (MEC) technology will be able to address many of the problems associated with this layer.

Data Storage Layer

Data storage units are contained in this layer, which stores raw data and information derived from physical system edge processing. The extra layer of security provides additional protection and can handle the massive amounts of data and traffic generated by potential applications.

Management Service Layer

This layer consists of sub-layers as follows:

- **Network Management** It involves changing the form of interaction between devices & data centers. Wireless Network Function virtualization (WNFV) is the most critical technology in this sub-layer. To improve the consistency of the IoT structure, the network topology will be updated at the same time as the communication protocols, such as 5G-IoT. Wireless Software Defined Networks WSDN is another useful technology for IoT network control and reconfiguration.
- **Cloud Computing** It is at this layer that edge computing data and information is being (re)processed in the cloud System, allowing the final transformed information to be extracted. Mobile devices can now perform this type of computing, known as Mobile cloud computing MCC in real-time. Therefore, parallel processing can occur across multiple devices, improving the IoT framework's scalability, power, and speed.
- **Data Analytics** The sub-layer utilizes new data analytics techniques to extract value from real-time data (manipulatable information). Any advancements in the Big Data algorithms will help this sub-layer process data more efficiently. With an increasing amount of data being collected in the

near future, the function of this sub layer will only become more significant as 5G and IoT closer together.

Application Layer

Here Software communicates with the above layers and data resides here, so it is not required to run at network speed. Vertical and Mobile Applications, Control technologies, and Business Intelligence and Analytics are all capable of revolutionizing vertical markets and business needs. Essentially, the Application layer allows business people to use data correctly at the right time to make the right decisions [64].

Collaborate and Processes Layer

Until it has an operation, the information it gathers from the previous layers and the data it receives is worthless. People use software and data to meet their specific needs. There are multiple users who can use the single app simultaneously for different purposes.

Security Layer

The purpose of this layer is to protect and cover all of the previous layers, but each section (where this layer intersects other layers) serves its own purpose. Architecture's security layer includes user authentication, data encryption, cloud security, and network access [9]. Additionally, the protection layer guards against and anticipates threats and cyber-attacks, using forensics to determine the type of attack and prevent it.

3.3.3.2 Autonomous Vehicles, Smart Cars

Rapid technological advances have allowed manufacturers to better meet customer demands and the manufacturing industry is undergoing dramatic change. Innovation and taking advantage of opportunities can be made possible by technology. The Internet of Things (IoT) will be the focus of Industry 4.0, which integrates artificial intelligence and data science. In order to monitor parts during manufacturing and assembly, sensors and tags are attached to them. In order to understand Industry 4.0's promise, Artificial Intelligence and Data Science are integrated with IoT. Sensors are also used to improve extend their lives, machine performance, predict when maintenance needs to be performed, and learn how machinery can be redesigned to have more energy efficiency.

The capabilities of self-driving cars will be greatly enhanced by 5G autonomous driving. Wireless connectivity systems, allowed by modern automotive advances, aim to make autonomous vehicles safer to drive. 802.11p

DSRC is there, it will take a considerable amount of money to install multiple access points. Several cellular carriers are launching applications that make autonomous driving possible, with LTE technologies bridging the gap between cellular vehicle-to-everything (C-V2X) and mainstream adoption of cellular vehicle-to-everything (C-V2X). LTE-V2X's latency does not meet the mission-critical V2V specifications, but it can be used as a stepping stone to lower-level ADAS functionality. In terms of meeting the rigorous specifications of mission-critical autonomous driving and ADAS systems, DSRC and 4G C-V2X (LTE-V) compete and complement each other [7], with each having different benefits and limitations.

3.3.3.3 *Manufacturing Sector and Smart Factory*

The launch of 5G has only sped up the production of the future factory. Connected devices in these futuristic factories can sense their surroundings and interact with one another, allowing for decentralized decision-making. Many people believe that this transition would depend on the capabilities of next-generation 5G networks. Sensors would likely be aplenty in tomorrow's smart factories [10], each tracking a different part of the working environment. This will include linked resources that use data such as position and accelerometer data to understand where and how they are being used and direct staff accordingly. Because of its high bandwidth, wireless flexibility, and low latency, 5G is an obvious option for manufacturers in these environments.

For real-time machine communication, ultra-reliable low latency communication is important. Use-cases that produce more data traffic and host a larger number of devices/sensors will benefit from increased bandwidth and support for higher device density. Network slicing enables virtual network isolation, which increases security and reliability. Mobile edge computing allows for the preservation of essential network features at the edge, improving flexibility and operational continuity.

Manufacturers can increase mobility by using more versatile factory layouts, enhance operations by enabling efficiency-boosting technologies like virtual reality and autonomous mobile robots, and unlock knowledge by turning data into actionable insights that boost productivity and sustainability [28]. A 5G network can also provide better communication for a wide network of sensors used in factory floor machines and robots for predictive maintenance. Adding flexibility and reducing costs would be easier and more cost-effective with improvements and alterations to factory floor plans. A 5G network, connectivity technology, and facility would benefit manufacturing, allowing it to change business models, revenue, and development. The automation of smart factories, the enhancement of internal and external communications, and the centralization of product development life cycle management are all attainable with one network.

3.3.3.4 Healthcare Industry

A. Robotic Tele-Surgery

It takes about 20 to 30 seconds for 4G LTE networks to respond. Telesurgery using advanced robotics is not suitable for latency. Although 5G can significantly enhance advanced telesurgery or robotic surgery because it improves low-latency connectivity, it can also lead to increased telesurgery costs. Doctors and robots would be able to connect seamlessly because of reliable networks and ultra-low latency. Doctors can handle surgical robots from afar in a smooth and flawless manner. Surgical robots can also perform a wide range of procedures in a fully automated clinical environment. Several hospitals and clinics have already used robotic telesurgery on both animals and humans with great success.

B. AR and VR

They plan to launch campaigns that will encourage healthcare providers to provide alternative therapies and minimally invasive treatments to patients suffering from chronic pain and anxiety. By providing targeted content, the 5G-powered AR and VR devices can engage and distract terminally ill patients. Via AR and VR experiences, the targeted content would increase the patient's comfort level.

C. Telemedicine

Telemedicine requires a network that supports high-definition, real-time video. Wired networks are used for this purpose. In the future, mobile networks will be able to manage remote appointments via telemedicine, allowing healthcare systems to expand the scope of telemedicine. The use of this technology will also enable healthcare systems to treat patients earlier and provide them with access to specialists. It can also make it easier for doctors and other staff members to collaborate.

D. AI-Driven Diagnosis

Artificial intelligence (AI) is used in new-generation medical diagnostic solutions to correctly classify rare items in blood, urine, and other body fluids. In addition, many health professionals use AI to conduct predictive diagnoses in order to detect health issues before patients notice symptoms. Without reviewing reliable and up-to-date patient data, AI cannot reliably diagnose health conditions. By enabling the collection and exchange of real-time data, 5G will allow clinical collaboration. Patients can continuously exchange real-time health data through wearable devices and internet-connected devices. By enhancing wireless network speed and reliability, 5G will allow hospitals and clinics to implement AI-driven predictive diagnosis. Real-time accelerated learning requires networks that can handle huge volumes of data reliably and at high bandwidth. Additionally, mobile devices demand data access from providers. With 5G networks, healthcare providers will be able to deliver the best care they can, no matter where they are within the hospital or clinic.

Derive value out of 5G networking technology

- Capturing all patient data in a reliable way.
- Rich, real-time medical information delivered on mobile devices.
- Increasing access to appropriate infrastructure.
- Greater data security and improved data storage.
- Ability to operate medical equipment remotely without delay.
- Integration of augmented and virtual reality technologies to enhance intern training.
- Support the use of smart medical devices and instruments through connectivity and operations.

3.3.3.5 *Smart Grids and Cities*

A. Smart Grids

An example of IoT application is the smart grid, which consists of electricity grid components such as power lines, transformers, and household electric meters all contain sensors and are coupled with communications devices in real time. The smart grid will help the electric company be more proactive about maintenance, handle distribution more effectively, and respond to outages more quickly. In order to enhance power grid security, reliability, and performance, smart grids combine conventional electricity systems with technology for controlling, communicating, and analyzing. The digitalization of power generation facilities improves performance and system efficiency. Nodes throughout the power infrastructure, including power generation, transmission, transformation, and distribution, are covered by smart Grid Communications systems [11]. The new digitized power generation System aims to increase the performance of power systems by providing high-speed, real-time, high-capacity, reliable, and stable communications networks. Availability of bandwidth used to transmit and receive data is significantly increased with 5G. Smart grids are capable of serving as an integrator and supporting a larger range of requirements. The 5G technology offers higher reliability than fiber optics and smaller-range wireless communication technologies and provides over-the-air access and disaster recovery. 5G is a perfect technology for smart and digital grids because of its wide-area seamless coverage, ultra-high latency, and roaming capabilities.

B. Smart Cities

In our technical world, 5G is a vital component in delivering improved networking. A smart city, for example, manages a city's various functions by integrating communication and information technologies and 5G networking solutions in a safe manner. Libraries, Schools, hospitals, power plants, transit systems, waste management, water supply networks, law enforcement, and other community programs are among those activities. There is a necessity to find a way to aggregate individual transactions, many layers of data, human movement, including traffic flows, security operation, energy consumption changes, and almost every other major component of contemporary economies. This aggregation could be supported by 5G technology. The amount of money saved by Smart Cities is amazing. Smart cities use the Internet of Things (IoT) to capture real-time data in order to better understand how demand trends are evolving and react with cost-effective and more efficient solutions. Digital city ecosystems, in general, are built on ICT architectures that link a variety of dedicated networks of sensors, connected vehicles, home appliances, mobile devices, communication gateways, and data centers [13]. According to IoT trends, the number of connected devices worldwide will hit 75 billion by 2025. The growing number of interconnected items produces a massive amount of data that the city can analyze locally to make better decisions about what improvements or new projects will help citizens the most.

3.4 A Vision and a Need for 6G Communications

The sixth-generation wireless networking infrastructure, also known as 6G, is the successor of 5G cellular technology. 6G networks are supposed to be able to use higher frequencies than 5G networks, allowing for higher data speeds and a much larger total capacity. A significant reduction in latency would almost certainly be needed. Sixth-generation wireless technology is currently being developed in support of cellular data networks [50].

The 5G will be replaced by this, and it will likely be much faster [49]. Most likely, the 6G networks will be broadband wireless networks with small coverage cells, just like their predecessors. Compared to the predecessor technologies,6G networks would be much more heterogeneous which supports technologies such as virtual and augmented reality (VR/AR) [8], ubiquitous instant messaging, widespread intelligence, and the internet of abilities that have yet to be seen in the mobile world. According to news, the groundwork for 6G has already been laid, and some analysts claim that 6G speeds could exceed 1 terabyte (TB) per second, which is 8,000 times faster than 5G.6G mobile technology is supposed to enable communications with a one-microsecond or even sub-microsecond latency, allowing communication almost instantaneous [34].

3.4.1 Development Projects

- **Electronics and Telecommunications Research Institute of South Korea:** It is easy to imagine that South Korea is far ahead of the curve, particularly when it comes to Terahertz band technology for the next generation of mobile communication. The 5G network will be five times faster than 6G, which is expected to have 100 times the speed of 4G LTE.
- **MIIT, China's Ministry of Industry and Information Technology:** China has invested heavily in technology in order to improve its 6G technology. Due to this, MIIT oversees and invests directly in the R&D process.
- **In Finland, the University of Oulu has created** 6Genesis, a six-generation research project. The project will last for eight years and will build concepts that will be compatible with 6G technology by 2040.
- **USA initiatives:** The United States is expected to make a 6G spectrum available for early research and development at frequencies between 95GHz and 3GHz; frequencies above 95GHz to 3GHz, however, need to receive approval from Federal Communications Commission.

IMT-2020 Standard International Mobile Telecommunications-2020 specifications for 5G networks were published by the International Telecommunication Union's Radiocommunication Sector (ITU-R) in 2015. 3GPP released R13 for 5G standards at the same time. The ITU is expected to complete 6G standardization ITU-R & IMT-2030 by the end of 2030, while 3GPP is expected to complete 6G standardization in R23.

In July 2018, the International Telecommunication Union (ITU) formed a focus workgroup to examine device technologies for B5G/6G systems. 6Genesis, a flagship program based on 6G technologies, was founded by the Academy of Finland in 2018.

3.4.2 System Architecture of 6G

Next-generation wireless networks will be able to support massive machine-type communication (mMTC) [45], owing to a large number of connected devices and access points (APs)/base stations (BSs). A coordinated multi-point (CoMP) transmission is created when various BS/APs support more than one device at the same time [39]. Huge devices can generate enormous quantities of data, necessitating extremely powerful processing units and secure transmission connections [22]. The backhauling connections could use optical fiber or photonic communications, and the central processing units could use ML and AI algorithms. Multiple relays or transmitters can be used in 6G communication systems, and the user can use diversity techniques to increase their SINRs (Signal-to-Noise-plus-Interference ratios), similar to virtual MIMO (multiple input multiple outputs) systems. BSs or APs in their vicinity will have an understanding of the features, locations, and quality of service specifications of end devices, and all such devices will be informed of the features, locations, and QoS specifications of the BSs or APs around them. It is possible to optimize the performance of the wireless network by using robust interference management/optimization strategies. A base-band processing unit will process data, which will be speedy enough to handle and transfer resources among a variety of end-users (BPUs). The air interface is the primary component that leads to significant advancements in wireless generations. 4G was overwhelmingly powered by Orthogonal Frequency Division Multiplexing (OFDM), rather than code division multiple access (CDMA) [24]. Similarly, the implementation of a new air interface would be critical to the design of the 6G device. Another critical part of the 6G system architecture is AI and machine learning. AI and machine learning will play a key role in 6G wireless systems self-healing, self-organization, and self-configuration. The 6G has also been pushed to introduce a new spectrum for connectivity due to spectrum congestion. As a result, this new spectrum will play an important role in the 6G system architecture. Because 6G will be able to support a wide range of devices, from IoT devices to live HD video stream transmission, it would need to be suitable for all existing technologies. Thus, in the 6G network, a scalable architecture using multiple radio access technologies (RATs) will be essential is shown in Figure 3.7.

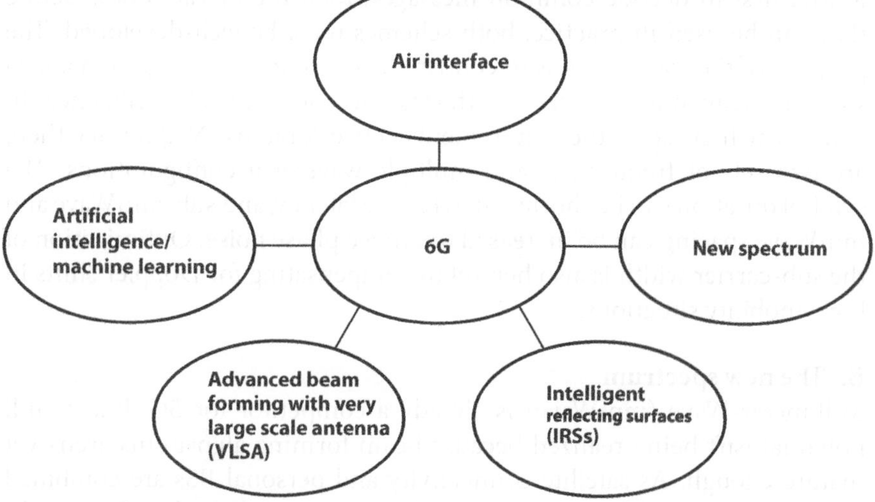

Figure 3.7 System's five key points of 6G.

A. Air Interface

6G focuses on the existing terahertz frequency spectrum, which has extremely wide bandwidths. It will present new challenges in terms of interacting efficiently at these frequencies [15, 16]. The solution would be to build a stable transmission system with sufficient range and low power consumption. The availability of extremely large bandwidths will shift the focus away from spectrally optimized solutions and toward solutions with better coverage. Making tradeoffs between spectrum accuracy, power consumption, and range will be critical in developing devices in these new frequency spectrums. This would entail the development of a new air interface in which single-carrier systems will be given more attention. OFDM must be reconsidered for lower frequency bands where spectral efficiency is important since the cyclic prefix only doubles the information and contains no additional information, and is energy-inefficient. Several researchers have suggested that a new scheme for a B5G/6G mobile network might be non-orthogonal multiple access NOMA.

In NOMA, all users have access to the same resource (frequency band) simultaneously. The rate-splitting multiple access (RSMA) technology has been proposed by some researchers as a potential access technology for 6G. Using successive interference cancellations (SIC), the user is able to decode the information in both NOMA and RSMA. The SICs are used by

RSMA first to decode common messages, then the private ones. Before they can be used in practice, both schemes must be well-developed. The purpose of the framework is to contribute to enhanced user QoS in a new software-defined air interface platform based on artificial intelligence. In 3GPP which discloses the requirements network radios (NR) for 5G, there are two sets of frequencies and multiple waveform configurations. The symbol duration can be shortened to reduce latency, and sub-mmWave and mmWave spacing can be increased to reduce phase noise. Optimization of the sub-carrier width is also helpful in compensating for Doppler shifts in high-mobility situations.

B. The new spectrum

Millimeter-Wave (mmWave) is already a competitor for 5G, but its full potential isn't being realized because beam forming algorithms aren't yet mature enough. As satellite connectivity and personal BSs are combined into cellular networks system, network upgrades are needed. Previously, the spectrum was split for a range of networks, including military communications, television (TV), and wireless communication [30, 33]. Due to this, it has been suggested to use an unlicensed spectrum (i.e., simultaneously using the THz band, mmWave, the visible spectrum of light) for wireless communication. As the signal travels over the higher frequency range, however, it attenuates rapidly. If a 3G/4G BS provides coverage over many miles, but a 5G/6G BS provides coverage over just a few hundred meters, then it is an inferior BS. For mmWave and THz communications, massive multiple input/multiple output concept (MIMO) and beam forming have been used.

C. Artificial Intelligence/Machine Learning

The 6G wireless communication will be at the center of society's digital transformation by providing secure, universal, and near-proximity instant messaging for humans and computers. In recent years, innovations in machine learning technology have enabled numerous new applications, such as self-driving cars [40]. As 5G/6G wireless networks become more sophisticated, it is necessary to employ more sophisticated methods to deal with any losses and detect network anomalies, features, and interpretation of KPI (Key Performance Indicator) patterns. This can be achieved by incorporating ML and SDN (Software Defined Networking) solutions. In order to achieve a certain level of KPI, ML/AI will be used to support the decision-making process. A new strategy is needed for the operation and implementation of RAN (Radio Access Network) for 6G [52]. Incorporating artificial intelligence into wireless algorithms (e.g., for CSI

(Channel State Information) feedback, channel estimation, and decoding, among other things) could result in a shift in the algorithms' direction. Utilizing algorithms with machine learning, artificial intelligence, and deep learning, we are able to efficiently handle resources based on user demand. Utilizing machine learning, deep learning, and artificial intelligence algorithms, we can immediately customize resources to meet user needs. This increases the likelihood of selecting the right solution, allowing the network to preserve its optimal state.

D. Very Large-Scale Antenna with Advanced Beam forming

The aim of beam forming is to direct the beam only in one direction. Considering that energy does not scatter evenly in all directions, focusing the beam in one direction increases the transmission range [47].

- **The Intelligent Reflecting Surfaces (IRS)**

The field of Intelligent Reflecting Surfaces (IRSs) may hold promise for beam forming in 6G. The material of an IRS can be made up of thin electromagnetic materials that intelligently alter the electromagnetic radiation route by changing the phase of the reflected rays using computer software. IRSs, for example, employ many low-power, cost-effective passive components to represent incident signals with user-configurable phase shifts without the need for encoding, decoding, increased power, demodulation, or modulation [35]. Signal-to-interference-plus-noise-ratio (SINR) can be improved using the IRS without requiring hardware or infrastructure changes to the network. Furthermore, no additional power is needed for the installation. IRS can reduce the number of antennas mounted at both the receiver and transmitter by reducing the number of antennae mounted there. As a result, the radio frequency (RF) chains at the receiver and transmitter will be reduced. IRS can substitute for traditional relay systems because of its benefits in terms of strength, spectral performances decreased hardware complexity. The IRS can be implemented during deep-fade or non-line-of-sight (NLOS) communication. Having and multiple IRS elements monitor the phases of the incident ray optimally at the receiver results in SINR being improved there. The IRS is unaffected by deterioration factor benefits such as noise and interference. IRS is a promising technology for 5G/6G communication systems due to these characteristics.

- **MIMO was made possible by orbital angular momentum (OAM)**

Having orbital angular momentum (OAM) of electromagnetic waves (EW) is a new dimensional property discovered in the late 1990s [42]. Multiple data streams may be transmitted over the same spatial channel as a result of this discovery. OAM is currently rated higher than other beam forming techniques due to its capacity to provide an enabling the EW to handle multiple data streams simultaneously over the same spatial channel, enhancing spectral efficiency, infinite number of orthogonal modes, and transmission power. OAM supports a large number of user (MDMA) schemes without using extra resources in mode division multiple entries [48] (i.e., frequency, time, and power). OAM offers a low probability of interception (LPI) option that can be implemented in a variety of frequency-hopping systems, including narrowband and wideband. The benefit of OAM-based MIMO systems over conventional MIMO systems is the capability and the long-range coverage (LOS). Therefore, 6G wireless networks are well suited for implementing OAM.

E. Coexistence of Variable Radio Access

A pervasive network will be able to be developed as a result of the coexistence of 6G flexible radio access technologies. In contrast, the nodes within this network are smart enough to determine the channel configuration and QoS requirements of each other. For instance, use cases and network coverage will determine the network type, such as cellular, Bluetooth, WLAN, and ultra-wideband (UWB), among others [37]. Due to this, the 6G communication standard must be built so that it converges with all other communication standards. Wi-Fi, UWB, Bluetooth, VLC (visible Light Communication), biosensors, UAVs, and satellite services will all be integrated into the 6G network and all come under the same norm so that they can all communicate with one another. Wi-Fi at 2.4 GHz has also made a significant impact on IoTs, with the majority of appliances already connected to it [52]. By integrating both of these technologies, 6G can utilize the vast infrastructure created by predecessor technologies that would otherwise cost it a lot of money [51]. Network densification, high throughput, high spectral performance, high reliability, low latency, and massive convergence are all characteristics of previous technologies that will be implemented in 6G. 6G technologies can also continue the trend of providing innovative services by incorporating new innovations such as VLC, AI/ML, QC, and quantum machine learning (QML). Smart houses, smart wearables, Smart vehicles, and 3D mapping are only a few of the services available.

3.4.3 Satellite Network of 6G

The 6G mobile system will be a combination of a 5G mobile wireless system and a satellite network for global coverage is shown in Figure

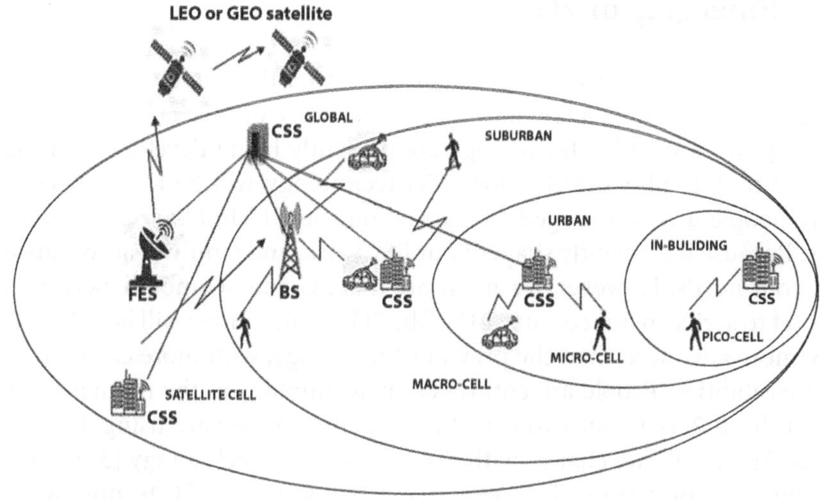

Figure 3.8 Satellite network of 6G.

3.8. A "satellite communications network," a "Earth-imaging satellite network," and a "navigation satellite network" are examples of satellite networks [43]. Global positioning system (GPS) satellites connect to telecommunications satellites that transmit voice, text, data, audio, Internet, and video; earth imaging satellites collect climate and environmental data, and satellite navigation systems transmit data related to global positioning. The United States GPS the four republics that created such satellite systems. China's Compass framework, the European Union's "Galileo system," and Russia's Glonass system [53]. The primary goal of 6G is to provide mobile phone users with a range of services such as a network identifier in a broad range of locations, multimedia apps, and access to the Internet for mobile users with fast data speeds without disrupting the network. It will give a lightning-fast Internet connection with incredible data rates of up to 11 gigabits per second. It provides a seamless wireless network with no boundaries. There is insufficient connectivity between network clients, resulting in Terabit-level transfer speeds [58]. Data and IOPS should be maximized (input and output per second operation).

3.5 Emerging of 7G

The majority of countries in the globe are currently employing 4G technology. However, only a few individuals are aware of this, and as a result, mobile phones with 7G technology are presently being developed. In the form of mobile phone technology, 7G technology will be a big success. Many people are encouraged to learn about new technology and inventions. Because it efficiently shapes your life, work, and family in accordance with your needs. However, we are aware that 5G and 6G mobile networks have yet to arrive in our country [31]. The 7G mobile phone will be substantially more advanced than the previous technology, with more capabilities and possibilities. People are enthralled by technology at the moment. We have utilized 2G-3G networks in the past, and now we are using 4G networks. As a result, we change with time, as does our technology [32]. We'll continue to improve ourselves over time. In the same way, 7G technology is now applied in technology. This technology will be far more advanced than any of its predecessors. 7G technology is the wave of the future.

The 7G mobile network which will be the most developed technology of the age, similar to the 6G network it will provide global coverage. This will be the most advanced generation of mobile communications. It will investigate demanding issues such as the need for mobile phones while moving from one country to other countries, because satellites are also moving at a constant speed and in specific orbit, from satellite-to-satellite correspondence frameworks and from cell to satellite framework principles and conventions. The fantasy of 7G can as it were to materialize when every one of the guidelines and conventions is characterized. Perhaps in the cutting edge, this is conceivable after 7G and named as 7.5G.

The difference between 6G and 7G is shown in Table 3.1. 7G networks will be able to utilize shorter communication latency, resulting in a significantly larger capacity and higher frequencies [36]. 7G will meet our requirements for extremely near-zero latency, high bandwidth, and universal integration. The 7G radio networks will provide communication and data collection capabilities, and are projected to enable significant advancements in imaging, data analytics, location awareness, Artificial General Intelligence (AGI), quantum computing, and high-performance computing.

By the time 7G networks are deployed, core and edge computing will be considerably more smoothly integrated as part of a combined computation/communications infrastructure framework [61]. With 7G technology going live, a range of potential benefits will become evident, including

Table 3.1 6G and 7G cellular network.

Parameters	7G	6G
Data Rate	About 11+Gbps	About 11Gbps
Multiplexing	CDMA	CDMA
Frequency Range	95GHz-3THz	95GHz-3THz
Switching type	All Packet	All Packet
Service Provided	Cellular services that are secure and global	Cellular services that are secure and global
Pros	Low-cost call, no issue of data coverage, and high data speed	Supports a higher number of connections compared to 5G, Offers very low latency, and a very high data rate
Cons	Drawbacks similar to 5G and 6G	Requires Perfect Scheduling, Limits on Flexible radio access

enhanced access to artificial general intelligence (AGI). Deep learning, Mind reading, AGI, Big data analytics, and other formerly independent technologies will be integrated and converged in the seventh generation of cellular wireless communications. The core communications network fabric will also change when many new technologies merge with 7G. It is important to note that AGI will be put front and center with 6G. Also, 7G will play a major role in public safety and thereby will enhance the health monitoring systems, Toxicity, sensing threat detection mechanisms, and IoT device management.

Only after all of 7G's standards and protocols are defined, the dream of 7G can come to reality. Perhaps this is possible only in the next generation after 7G, which can be called 7.5G or 8G. When 7G has addressed all of its flaws, there will be no issues with data capacity coverage or hand-off. There would be only one demand from users at that time, which will be the fee of a mobile phone call and associated services [54]. This issue will usher in a new era of standardization and technology, as well as open up new horizons for computing research. 7.5G or 8G will be the next technology for lowering the cost of mobile phone calls and services. The peak download and upload data rates of Generation 7.5 are extremely fast. To view the high-quality video in this location, time block codes are used for transmission. We can download five movies with a total length of 20 minutes, GB files in a second and upload 1 GB files or any other type of file data. As a result, it navigates to the satellite network. FEC and the OFDM approach are both used to speed up the process, accelerating the communication process.

- **Features and Capabilities of 7G**

7G is proposed to combine with a Satellite network for worldwide coverage. It aims to deliver space roaming at extremely high data rates using existing satellites such as those used for imaging of earth and navigation, telecommunications, and also for other purposes. The navigational satellite will be used for global positioning systems, while the telecoms satellite will be used for multimedia and voice communications [56]. The earth-image satellites will give us real-time weather updates at a faster pace as well as assistance with natural catastrophe preparedness. 7G will serve as the foundation for a fully linked society with high-speed Internet, low-cost, wireless speeds of up to 11Gbps, and the potential to tap satellite communication networks via specially developed nano antennas [55]. 7G technology is expected to provide AI-powered network management, and signal transmission, continuous computing, Remote access for learning, diagnosis, high-speed data, and Virtual area surroundings with realistic sensations.

Findings

- In comparison to its predecessors, the seventh generation (7G) of wireless mobile communication will be the most advanced. It will provide well-defined satellite communication standards and protocols, as well as improve security. For the speed of the communication process, 7G systems can be supported by global navigation satellite systems using techniques such as OFDM methodology and FEC.
- It will improve the service quality (QoS). Include services like direct HD video transmission to improve 7G. It must achieve "Zero Latency," which previous generations have failed to do. It must be a low-cost technology with no problems with data capacity coverage or hand-off. It must set a goal of providing a network for everyone and implementing net neutrality i.e providing constant and equal speed to all users.
- 7G has developed a Tier-1 network that serves carriers, call centers, and companies. 7G Network uses cutting-edge

architecture and technology to reliably deliver millions of calls every day, all over the world, at industry-leading rates. Customer portal features provide access to service analytics and reporting, customer account and service modifications, and options for shifting more of the services to a trusted network. The reliability of the 7G Network is built on the service uptime, network hardware redundancy within the Network Operation Center. This provides the 7G network to be available during natural disasters.

3.6 Conclusions

In the next few years, 5G services and networks will be introduced in phases to provide a forum for emerging digital services and business models to flourish. 5G would be a watershed moment in telecom history, bringing fast access to an unlimited number of devices. It will allow machines to communicate in an IoT environment that can power an almost unlimited number of services. As IoT use cases multiply and more devices become connected, 5G networks will allow the Internet of Things to grow rapidly and will be beneficial to businesses and consumers 5G networks will revolutionize transport and provide better access to medical care by reliably tying together patients and physicians worldwide. As the user interface shifts away from the text, images, and videos and toward virtual reality and augmented reality, 5G cellular service will help to promote this new change by providing fast speeds, reliable service, and extremely large bandwidth, and also reducing latency. The importance of 5G in the performance of various industries, such as the Internet of Things, smart vehicles, manufacturing and smart factories, smart grids, and smart cities, and healthcare were also discussed. It addressed 5G is crucial for increasing market automation and addressing the multiple issues that various manufacturing industries are facing in today's rapidly evolving environment. In conclusion, the transition to 6G communications and the architecture of the 6G framework has been discussed. As the world attempts to become entirely wireless, demands are growing for continuous access to information from any location at any time, with fast speed, better quality, and expanded connectivity. 5G follows 4G, and 5G along with satellite broadband is expected to become a 6G network. However, because of varying standards and technologies, handoff/roaming will be problematic. As a result, the 7th generation of wireless mobile networks attempts to achieve space roaming.

References

1. Ian F. Akyildiz, Shuai Nie, Shih-Chun Lin, Manoj Chandrasekaran, *5G roadmap: 10 key enabling technologies*, http://dx.doi.org/10.1016/j.comnet.2016.06.010 ISSN 1389-1286/© 2016 Elsevier B.V. All rights reserved Computer Networks Volume 106, 4 September 2016, Pages 17-48.
2. I.F. Akyildiz, D.M. Gutierrez-Estevez, R. Balakrishnan, E. Chavarria-Reyes, *LTE-Advanced and the evolution to Beyond 4G (B4G) systems*, http://dx.doi.org/10.1016/j.phycom.2013.11.009. *Phys. Commun.* 10(2014) 31–60.
3. Shih-Chun Lin, Manoj Chandrasekaran *3GPP TS* 25.913, *Requirements for evolved UTRA and evolved UTRAN*, 2009.
4. Hamed Rahimi; Ali Zibaeenejad; Ali Akbar Safavi *A Novel IoT Architecture based on 5G-IoT and Next Generation Technologies* 2018 IEEE 9th Annual Information Technology, Electronics and Mobile Communication Conference (IEMCON).
5. Shaikh Abdul Azeem, Satyendra Kumar Sharma *5G Network and Issues Related to Network Convergence, International Journal of Current Advanced Research* ISSN: O: 2319-6475, ISSN: P: 2319-6505, Available Online at www.journalijcar.org DOI: http://dx.doi.org/10.24327/ijcar.2018 Volume 7; Issue 5(B); May 2018.
6. Dongmei Yang , Yunhui Zhou, Wentao Huang, Xianwei Zhou *5G mobile communication convergence protocol architecture and key technologies in satellite internet of things system*, https://doi.org/10.1016/j.aej.2020.09.019 1110-0168 2020 Faculty of Engineering, *Alexandria University. Production and hosting by Elsevier B.V.*
7. T. Xia, M.M. Wang, J. Zhang, *et al*, *Maritime internet of things: challenges and solutions, IEEE Wirel. Commun.* 27 (2) (2020) 188–196.
8. L. Zhang, Y.C. Liang, D. Niyato, *6G Visions: Mobile ultra-broadband, super internet-of-things, and artificial intelligence*, China Commun. 16 (8) (2019) 1–14.
9. C.K. Wu, K.F. Tsang, Y. Liu, *et al.*, *Critical Internet of Things: An Interworking Solution to Improve Service Reliability, IEEE Commun. Mag.* 58 (1) (2020) 74–79.
10. Mohsen Attaran "*The impact of 5G on the evolution of intelligent automation and industry digitization*" https://doi.org/10.1007/s12652-020-02521-x *Journal of Ambient Intelligence and Humanized Computing.*
11. Raúl Chávez-Santiago Michał Szydełko Adrian Kliks Fotis Foukalas Yoram Haddad Keith E. Nolan Mark Y. Kelly *5G: The Convergence of Wireless Communications* Moshe T. DOI 10.1007/s11277-015-2467-2 Masonta Ilangko Balasingham Wireless Pers Commun (2015) 83:1617–1642
12. Agyapong, P., Iwamura, M., Staehle, D., Kiess, W., & Benjebbour, A. (2014). *Design considerations for a 5G network architecture*, doi:10.1109/MCOM.2014.6957145, *IEEE Communications Magazine*, 52(11), 65–75.

13. Aijaz, A., Aghvami, H., & Amani, M. (2013). *A survey on mobile data off-loading: Technical and business perspectives. IEEE Wireless Communications,* 20(2), 104–112.
14. Akyildiz, I. F., Gutierrez-Estevez, D. M., Balakrishnan, R., & Chavarria-Reyes, E. (2014). *LTE-advanced and the evolution to beyond 4G (B4G) systems. Physical Communication,* 10, 31–60. doi:10.1016/j.phycom.2013.11.009 Physical Communication Volume 10, March 2014, Pages 31-60.
15. Faiza Nawaz, Jawwad Ibrahim, Maida Junaid, Sabila, Muhammad Awais ali A Review of Vision and Challenges of 6G Technology (IJACSA) International Journal of Advanced Computer Science and Applications, Vol. 11, No. 2, 2020.
16. Zhao, Y., Yu, G., & Xu, H. (2019). 6G Mobile Communication Network: Vision, Challenges and Key Technologies. arXiv preprint arXiv:1905.04983 SCIENTIA SINICA Informationis, Volume 49, Issue 8: 963-987(2019).
17. David, K., & Berndt, H. (2018). 6G vision and requirements: Is there any need for beyond 5G? IEEE vehicular technology magazine, 13(3), 72-80.
18. Cacciapuoti, A. S., Sankhe, K., Caleffi, M., & Chowdhury, K. R. (2018). Beyond 5G: THz-based medium access protocol for mobile heterogeneous networks. IEEE Communications Magazine, 56(6), 110-115.
19. Calvanese Strinati, E., Mueck, M., Clemente, A., Kim, J., Noh, G., Chung, H., ... & Destino, G. (2018). 5G CHAMPION-Disruptive 5G Technologies for Roll Out in 2018. ETRI Journal, 40(1), 10-25.
20. Yadav, Swati, and Sugandha Singh. "Review paper on development of mobile wireless technologies (1G to 5G)." Int. J. Comput. Sci. Mob. Comput 7.5 (2018): 94-100.
21. Guangyi Liu, Dajie Jiang, "5G: Vision and Requirements for Mobile Communication System towards Year 2020", https://doi.org/10.1155/2016/5974586, Chinese Journal of Engineering, vol. 2016, Article ID 5974586, 8 pages, 2016.
22. Zheng, Liping. (2018). 5G Mobile Communication Development Trend and Several Key Technologies. MATEC Web of Conferences. 246. 03034. 10.1051/matecconf/201824603034.
23. Chen Jiaming. Key technical specifications and development of 5G mobile communication [J]. Telecommunications Engineering Technology and Standardization, https://doi.org/10.1051/matecconf/201824603034 ISWSO 2018. MATEC Web of Conferences 246, 030 (2018)
24. Wenxuan Long, Rui Chen, Marco Moretti, Wei Zhang, Jiandong Li "A Promising Technology for 6G Wireless Networks: Intelligent Reflecting Surface", Journal of Communications and Information Networks, Vol.6, No.1, Mar. 2021
25. Cacciapuoti, A. S., Sankhe, K., Caleffi, M., & Chowdhury, K. R. (2018). Beyond 5G: THz-based medium access protocol for mobile heterogeneous networks. IEEE Communications Magazine, 56(6), 110-115.

26. Calvanese Strinati, E., Mueck, M., Clemente, A., Kim, J., Noh, G., Chung, H., & Destino, G. (2018). 5G CHAMPION–Disruptive 5G Technologies for Roll-Out in 2018. ETRI Journal, 40(1), 10-25.
27. Rupayali Swaroop and Arun Kumar" An Overview & Evolution of 5G Wireless Mobile Technology" International Journal of Information Technology and Electrical Engineering, Volume 9, Issue 6 December 2020.
28. Mamta Agiwal, Navrati Saxena & Abhishek Roy (2019) Towards Connected Living: 5G Enabled Internet of Things (IoT), IETE Technical Review, 36:2, 190-202, DOI: 10.1080/02564602.2018.1444516.
29. Sahoo, Ashok & Goyal, Parul. (2019). A Roadmap Towards Connected Living: 5G Mobile Technology. International Journal of Innovative Technology and Exploring Engineering. 9. 2278-3075. 10.35940/ijitee.A4742.119119.
30. Dr.Hridaya Nand Sah. "A Brief History of Mobile Generations And Satellite Wireless Communication System." 2017 Jetir December 2017, Volume 4, Issue 12 Www.Jetir.Org (Issn-2349-5162)
31. T. Arunkumar and L. Kalaiselvi, ——Latest Technology of Mobile Communication and Future Scope of 7.5 G||, International Journal of Engineering & Technology Research, Volume 2, Issue 4, pp. 23-31, July 2014.
32. Gawas, A. U. (2015). An overview on evolution of mobile wireless communication networks: 1G-6G. International Journal on Recent and Innovation Trends in Computing and Communication, 3(5), 3130-3133.
33. Gill, J., & Singh, S. (2015). Future prospects of wireless generations in mobile communication. Asian J. Comp. Sci. Technol, 4(2), 18-22.
34. Zong, B., Fan, C., Wang, X., Duan, X., Wang, B., & Wang, J. (2019). 6G Technologies: Key Drivers, Core Requirements, System Architectures, and Enabling Technologies. IEEE Vehicular Technology Magazine, 14(3), 18-27.
35. Yang, P., Xiao, Y., Xiao, M., & Li, S. (2019). 6g wireless communications: Vision and potential techniques. IEEE Network, 33(4), 70-75.
36. Khutey, R., Rana, G., Dewangan, V., Tiwari, A., & Dewamngan, A. (2015). Future of wireless technology 6G & 7G. International Journal of Electrical and Electronics Research, 3(2), 583-585.
37. Kalbande, D., Haji, S., & Haji, R. (2019, June). 6G-Next Gen Mobile Wireless Communication Approach. In 2019 3rd International conference on Electronics, Communication and Aerospace Technology (ICECA) (pp. 1-6). IEEE.
38. Bastug, E., Bennis, M., Médard, M., & Debbah, M. (2017). Toward interconnected virtual reality: Opportunities, challenges, and enablers. IEEE Communications Magazine, 55(6), 110-117.
39. Yajun, Z., Guanghui, Y., & Hanqing, X. U. (2019). 6G mobile communication networks: vision, challenges, and key technologies. SCIENTIA SINICA Informationis, 49(8), 963-987. [16] A. Checko *et al.*, ——Cloud RAN for mobile networks—A technology overview, IEEE Commun. Surveys Tut., vol. 17, no. 1, pp. 405–426, 2015. doi: 10.1109/COMST.2014.2355255.

40. Al-Eryani, Y., & Hossain, E. (2019). The D-OMA Method for Massive Multiple Access in 6G: Performance, Security, and Challenges. IEEE Vehicular Technology Magazine, 14(3), 92-99.
41. B. Li, Z. Fei, and Y. Zhang, —UAV communications for 5G and beyond: recent advances and future trends, || IEEE Internet of Things Journal, vol. 6, no. 2, pp. 2241-2263, April 2019.
42. Chowdhury, M. Z., Shahjalal, M., Ahmed, S., & Jang, Y. M. (2019). 6G Wireless Communication Systems: Applications, Requirements, Technologies, Challenges, and Research Directions. arXiv preprint arXiv:1909.11315. (2019). 6G. [Online]. Available: http://mmwave.dei.unipd.it/research/6g/
43. Dang, S., Amin, O., Shihada, B., & Alouini, M. S. (2019)." From a Human-Centric Perspective: What Might 6G Be?". arXiv preprint arXiv:1906.00741. Nature Electronics, 2020 vol:3 ISSN :2520-1131
44. Sanjeev Singh, Rakesh Kumar Jha "A Survey on Software Defined Networking: Architecture for Next Generation Network" Issue 2/2017
45. Mohsen Gerami "Evaluation of Next Generation Networks "(IJCSE) International Journal on Computer Science and Engineering Vol. 02, No. 02, 2010, 378-381.
46. Satya N Gupta ,Next Generation Networks (NGN)-Future of telecommunication, International Journal of ICT and Management, February 2013 vol-I Issue-I,ISSN NO.2026-6839.
47. W. Roh, J.-Y. Seol, J. Park *et al.*, "Millimeter-wave beamforming as an enabling technology for 5G cellular communications: theoretical feasibility and prototype results," IEEE Communications Magazine, vol. 52, no. 2, pp. 106–113, 2014.
48. E. G. Larsson, O. Edfors, F. Tufvesson, and T. L. Marzetta, "Massive MIMO for next generation wireless systems," IEEE Communications Magazine, vol. 52, no. 2, pp. 186–195, 2014.
49. H. Nikopour, E. Yi, A. Bayesteh *et al.*, "SCMA for downlink multiple access of 5G wireless networks," in Proceedings of the IEEE Global Communications Conference (GLOBECOM '14), pp. 3940–3945, Austin, Tex, USA, December 2014.
50. B. Li, H. Shen, D. Tse, and W. Tong, "Low-latency polar codes via hybrid decoding," in Proceedings of the 8th International Symposium on Turbo Codes and Iterative Information Processing (ISTC '14), IEEE, Bremen, Germany, August 2014, pp. 223–227.
51. C. Zhang and K. Parhi, "Low-latency sequential and overlapped architectures for successive cancellation polar decoder," IEEE Transactions on Signal Processing, vol. 61, no. 10, pp. 2429–2441, 2013.
52. B. Yuan and K. K. Parhi, "Low-latency successive-cancellation polar decoder architectures using 2-bit decoding," IEEE Transactions on Circuits and Systems I: Regular Papers, vol. 61, no. 4, pp. 1241–1254, 2014.

53. Shuji Esaki, Akira Kurokawa, and Kimihide Matsumoto, 2007, Overview of the Next Generation Network, NTT Technical Review, Vol. 5 No. 6 June 2007.
54. IMT-2020 (5G) Promotion Group, "5G vision and requirements," http://www.imt-2020.cn.
55. B. Bhattacharyya and S. Bhattacharya, Emerging Fields in 4G Technology, its Applications & Beyond-An Overview, International Journal of Information and Computation Technology. vol. 3, no 4, pp. 251-260, 2013.
56. S. Mondal, A. Sinha and J.Routh, A Survey on Evolution of Wireless Generations 0G to 7G, International Journal of Advance Research in Science and Engineering- IJARSE, Vol. 1, Issue 2, pp. 5-10.
57. Ashish Kr. Gupta, Madan Pal Singh N. "A Study of Next Generation Wireless Network 6G", International Journal of Innovative Research in Computer and Communication Engineering, 4(1), 871-874 2018.
58. Zhang, Y., Wang, F., Wang, D. & Guo, D. (2020). Slicing Resource Allocation for eMBB And URLLC in 5G RAN, Wireless Communications and Mobile Computing, 2020, 1-11.
59. Samaresh, B., Sudip, M. & Athanasios, V.V. (2017). Software-Defined Networking for Internet of Things: A Survey, IEEE Internet of Things Journal, 4(6), 1994-2008.
60. Cengiz, K. (2018). Next-Generation Infrastructure and Technology Issues in 5G And Systems, 14(1), 33-39.
61. Arvidsson, A. & Westberg, L. (2019). Transport Bottlenecks of Edge Computing in 5G Networks,Journal of Communications Software And Systems,15(1), 59-65.
62. Wang, C., Haider, F., Gao, X., You, X., Yang,Y., Yuan, D., Aggoune, H., Haas, H., Fletcher, S. & Hepsaydir, E. (2014). Cellular Architecture and Key Technologies for 5G Wireless Communication Networks, IEEE Communications Magazine, 52(2), 122-130.
63. Yunzheng, T., Long, l., Shang, L., & Zhi, Z. (2016). A Survey: Several Technologies of Nonorthogonal Transmission for 5G, Wireless Communication over ZigBee for Automotive Inclination Measurement, 12(10), 1-15.
64. Vilceanu, C., Grecea, C. & Herban, S. (2017). Spatial data geoportal for Local Administration–smart solution for a secure and valuable cultural heritage, Journal of Geodesy, Cartography and Cadastre, 7(1), 12-18.
65. Kashaf, A., Tiwana, M.I., Usman, I. & Tiwana, M.I. (2017). Self-organizing inter-cell interference coordination in 4G and beyond networks using genetic algorithms, Automatika, 58(1), 48-54.

4

Understanding the Salient Features Related To Resource Management in Broadband Wireless Networks

Ramkumar Jayaraman[1]*, Devandar Rao[1], Manoj Kumar[2] and Ashutosh Mishra[3]

[1]Department of Computing Technologies, School of Computing, SRM Institute of Science and Technology, KTR Campus, Chennai, India
[2]University of Wollongong, Faculty of Engineering and Information Sciences, Dubai, UAE
[3]School of Integrated Technology, Yonsei University, Seoul, South Korea

Abstract

In recent days, Wireless technologies are improving drastically based on the number of users, leading to network traffic and reliability in the data services. Here the resource plays a significant role, and it needs various approaches to manage and monitor those resource mesh networks. As the network is concerned, different structure/topology is applied as the framework, and mesh topology is used in recent networking technologies. The various scheduling approaches are discussed based on the vast resource in the network. Then different characteristics are discussed related to radio channels based on wireless broadband networks and to provide effective network traffic management and Quality of Service. As advanced networks is concern, network coding approaches are considered while the data are exchanged between the nodes in the network. Finally, the various parameters are discussed related to resource and channel management to provide effective communication in the network.

***Keywords*:** Network traffic, wireless technology, interference, topology, resource management, quality of service

**Corresponding author*: ram.kumar537@gmail.com

A. Suresh, J. Ramkumar, M. Baskar and Ali Kashif Bashir (eds.) Resource Management in Advanced Wireless Networks, (81–98) © 2023 Scrivener Publishing LLC

4.1 Introduction

As advanced wireless network is concerned, it mainly focuses on high-speed data services to access the internet across the wide area region. As the network is concern, the user initiates certain services and it is provided by the network provider. In the wireless network, Base Station (BS) act as transceiver, which data can be send/receive to manage the services in the network. BS provides range in terms of radio coverage and it cover over large geographical area. It has several Subscriber Stations (SS) attached to it and it provides the services to all the mobile user in that radio coverage area [1]. By using BS, mobile users are initiated with certain services and subscriber equipment are listed as, mobile phones, laptop, etc., and communication gets established between the mobile devices via base station as represented in Figure 4.1.

As the network is generally concern, there are two types of connections are possible as Transport Control Protocol (TCP) [2] and User Datagram Protocol (UDP) [3], which is basically referred as connection and connectionless oriented communication. As the network is concern, the IEEE is the standard, which assign and manage the prototype standard to each and every wireless network technology. For the broadband wireless network, IEEE 802.16 is referred and it is nothing but the fixed network [4].

It has certain advantages and disadvantages related to seamless connection establishment and mobility feature. In this networking aspects, resource plays a major role in the service management, which is initiated by the network. For effective management of resource, the scheduling and

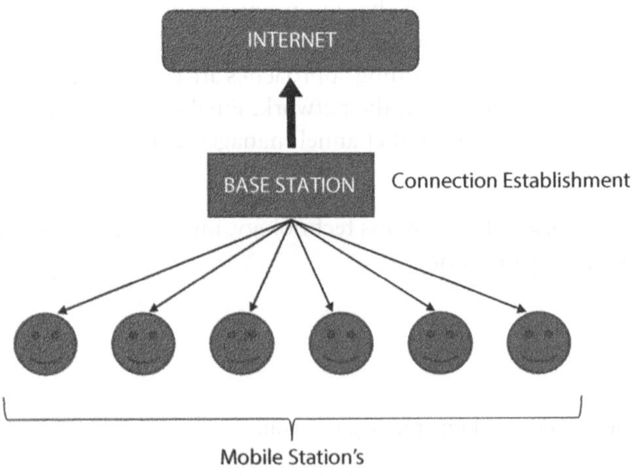

Figure 4.1 Wireless networks structure.

Table 4.1 Network technologies comparison with respect to the salient features.

Comparison	Year	Technology	Access system	Bandwidth
2G	1993	GSM	TDMA, CDMA	25Mhz
3G	2001	WCDMA	CDMA	25Mhz
4G	2009	LTE, WiMAX	CDMA	100Mhz
5G	2018	MIMO, mmWaves	OFDM, BDMA	300Ghz

proper managing of resources are to be done in the wireless networks. This scheduling, managing the resource and Quality of Service (QoS) comes under the layer of Medium Access Control (MAC) [5]. The growth of the networking technologies is compared based on the mobility, network range and network bandwidth in terms of Km/h, Km and Mbps/Gbps and it is tabulated in Table 4.1.

As IEEE 802.16 gets enhanced from a to m and soon, IEEE 802.16m [6] is denoted as mobile WiMAX as it supports network mobility as it attains the bit rate up to 100 Mbits/sec and it can covers the network range of 5Km as the mobile stations can be utilized with certain services. In this standard, BS provides effective radio coverage to initiate the connection establishment with all the MS in that coverage with full mobility support as BS is connected to the internet directly.

In the wireless networks, Mobile WiMAX has certain salient features as mentioned below in terms of data rate, Quality of Service and Mobility. The network service is more flexible in managing the data in terms of reliability based on the coding schemes. The QoS to be maintained with helps of TDMA through air interface to determine the optimal scheduling with effective QoS [7] and [8]. While performing handover either soft/hard to perform effective mobility to obtain proper Voice over Internet Protocol (VoIP).

4.1.1 Network Topology Related to Wireless Networks

As network topology is a concern, generally there are several network topologies related to networks. While specific to wireless networks, there are two network topologies are mentioned below,

a. Point to Multi-Point (PMP) Network Topology:
   Based on this network topology related to wireless networks [9], there is only one BS is used, which acts as a central controller/entity and it performs managing and monitoring

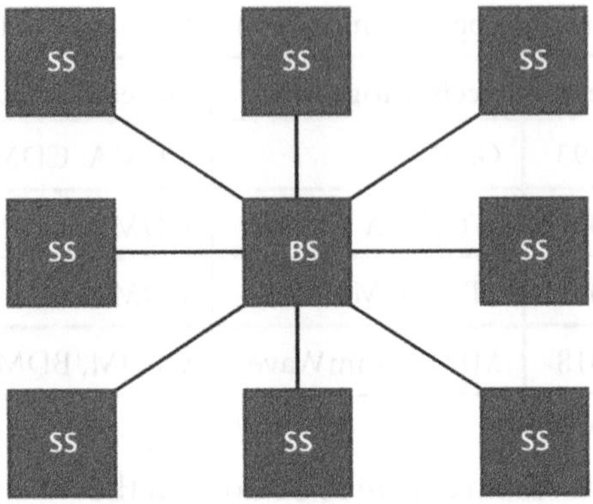

Figure 4.2 PMP network topology.

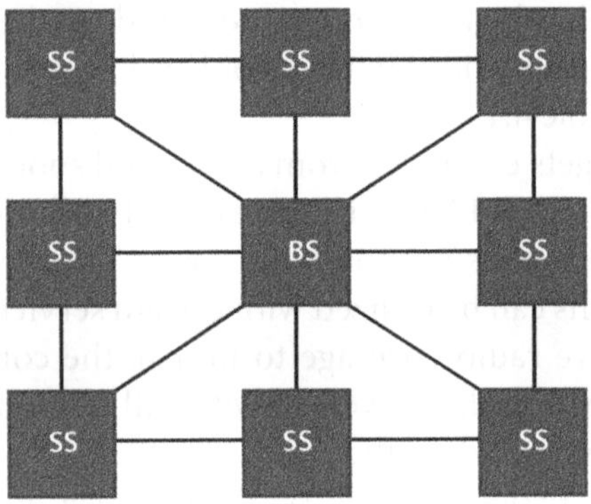

Figure 4.3 Mesh network topology.

all the MS connected to this BS networks as represented in Figure 4.2.

b. Mesh Network Topology:
Based on this network topology related to wireless networks [10], there is a BS connected to all the SS and it is organised in a distributed manner. So that each SS are connected with each other and it can exchange the packets and route to each other by forwarding. It supports multi-hop communication feature as represented in Figure 4.3.

4.1.2 Challenges in Wireless Broadband Network

There are several technical challenges are used and it helps to deploy an effective wireless broadband network. As wireless broadband network is concern, it delivers certain services related to QoS such as, throughput, bandwidth, delay, etc., which provide services at the speed in terms of megabits/sec to all thesenduserwhoareutilizingtheservicesforbothdata,voiceandmultimedia [11]. As there is a significant development in the internet era, IP versions are getting developed with respect to the real time applications and provide the services effectively [12]. As the indoor internet is concern, the user can deploy by their own internet deployment to utilize the services with some handheld devices. Then after the indoor devices, mobile based wireless networks will provide the services at high speed with continuous connection to certain devices like handheld, laptop, etc [13]. Now the users in the need of certain services with compromising the quality, security and reliability.

There some technical challenges are listed below,

- Through wireless communication, effective data services are performed based on data transmission reliably and data reception approach is needed.
- As there are limited number of frequency spectrum, high efficiency spectral resource and radio coverage make the services to be delivered effectively.
- Effective multiplex services related to various QoS services such as, throughput, delay, packet loss, etc. are enabled.
- Enabling the wireless mobility in dynamic fashion makes services like handover and roaming continuously.
- As low power devices are used, low power consumption can be obtained.
- Robust security.
- Low cost and operational complexity are considered by applying IP based approaches and framework for wired/wireless network is needed.

4.1.3 Radio Channel

Generally wireless broadband network gets initiated based on the communication medium as it is classified based on wired and wireless network. In wired networks, communication can be established based on some physical elements such as, cables, copper wired and fibre connection, which helps to transmit/receive the signals to establish communication [14]. In wireless networks, complex communication is considered based on some wave propagation approaches to mitigate the intervention signals while considering the communication [15]. There are two types of sight is available as

- Line of Sight (LoS)
- Non-Line of Sight (NLoS)

Intervention occurs when there is while transmiting/receiving the data in network. Related to intervention, radio wave propagation is needed, as it is a complex structure for wireless communication systems. As broadband wireless networks are concerned, signals are to be transmitted under the No Line of Sight (NLoS) action. When there is an obstacle, small or big, during the signal transmission from one medium to another, interference or another problem can occur. There are certain other factors like signal delay, signal strength, and distortion, as it may vary from time to time. Mainly the challenges are considered based on the design of digital wireless communication using high data rate and speed mobility, as represented in Figures 4.4 and 4.5.

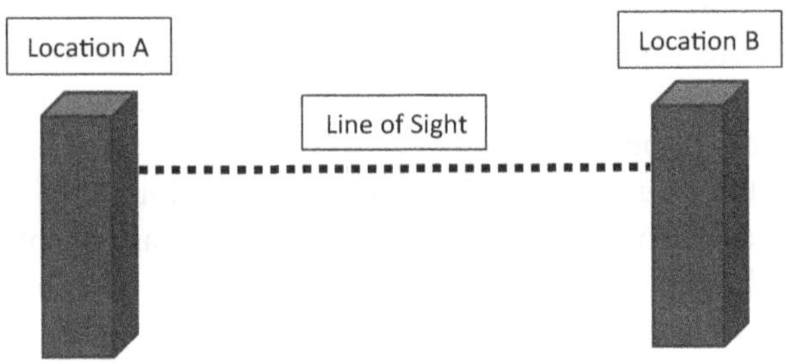

Figure 4.4 Line of Sight (LoS).

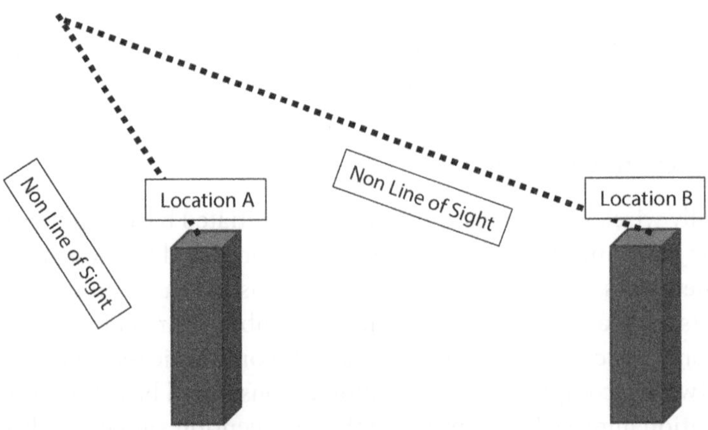

Figure 4.5 Non Line of Sight (NLoS).

When there is a decrease in transmitting the signal power as it depends on the longer distance. When NLoS is concern, the signal to be received at the receiver end gets decrease the power signal as distance is long. When LoS is considered, signal strength is much higher when compared to the signal on NLoS. Based on the distance scenario, pathloss gets calculated using the attributes of system terrain, obstacles and height of antenna.

When there is a block in the signal due to large obstructs in between the signal transmission. There is a loss in the power signal, when there is a wave of radio signals gets propagates and blockage occurs. Based on the distance, the signal strength gets degraded and pathloss occurs [16]. Received signals has longer variation as there is signals are reflected and based on the objects, signals also get scattered, when the transmission signals get transmitted to the received signals through various path or channel to flow. When the above signals are flowed at various path leads to multipath fading problem and it is termed by the variation in amplitude. Due to the large variation based on the strength of the received signals, it leads to the wireless broadband system design.

Based on the dispersion in the time as inter symbol interference occurs in the multipath environment. The inter symbol interference occurs as there is a delay in the time based on the signal path significance as the transmitted signal time get some delay variation as the it is gets received by the receiver. When there is a lower signal delay, data rate will be much higher and lower cause in the inter symbol interference problem. As the above problem is the challenging task when considering the wireless broadband networks and to be mitigated. While considering the existing methods, equalization is the concept to be in cooperated to overcome ISI problem to achieve data rate with power processing. When considering WIMAX technologies, (Orthogonal Frequency Division Multiple Access (OFDMA) is used to mitigate the ISI problem.

During the mobility feature, frequent power dispersion occurs. Here the concept of Doppler Effect is added and it is stated as motion occurs between the transmitter and receiver and it causes frequency dispersion and as it related to speed of the vehicle and frequency carry. For the advanced wireless broadband technologies, Doppler Effect leads to loss in the Signal to Noise Ratio (SNR) value and this results in making the process of recovery and synchronization to be more difficult. When considering the OFDM, Doppler Effect makes OFDMA [17] sub carrier to get corrupted. Regarding the noise during the communication channel, Additive White Gaussian Noise (AWGN) is added into it as the bandwidth is proportional to the receiver with thermal noise present. By considering the standard narrow systems, receiver with noise value based on the broadband wireless networks is determined and it makes increase path loss to minimize the range of the coverage [18].

The interference problem occurs, when there is a limitation based on the spectrum availability whenever the user is sharing the bandwidth. When the users share the signals from various number of users results in interference problem and it results in large variation related to noise are addressed over here.

4.1.4 Scarcity in the Spectrum Utilization

The problem of scarcity in the radio spectrum utilization plays a significant role while addressing the broadband wireless networks. There are some limited spectrum can be utilized for the resource sharing as it is non-commercial use. Limitation of spectrum usage happens when there is an increase in the enormous number of users and as they requires high bandwidth. The current cellular architecture requires to sort out the problem of spectrum limitation. As the cellular network replaces the usage of higher transmitter as it requires high power to cover large area but there uses the multiple number of small transmitter as it cover larger area as it requires small power. Small transmitter refers as the cell as it denotes the antenna of directional region and here the multiple number of cells are grouped into cluster and it requires lesser number of spectrum availability. It helps to mitigate the interference formation among the cells in networks.

Spectrum utilization happens based on the allocation of resource repeatedly in many areas through the concept of reuse. While considering parameter of capacity and spectral improvement, reuse of the channel frequency can perform maximized reuse. Whenever the reuse gets increase and it is directly proportional to the increase in the channel inference, which leads to the network performance degradation. To eradicate the network degrade, the reuse should be performed tightly as the challenges of applying Signal to Interference plus Noise Ratio (SINR) condition on transceiver scheme in order to provide effective interference method. So, multiple antenna process is applied to provide effective interference method. Other than multiple antenna processing, there are several processing antenna methods are used to obtain the improved effective channel spectrum utilization and those information related to channel are exploited in order to improve the channel capacity as represented in Figure 4.6(a) and (b).

There are certain techniques are needed for the above performance improvement as listed below,

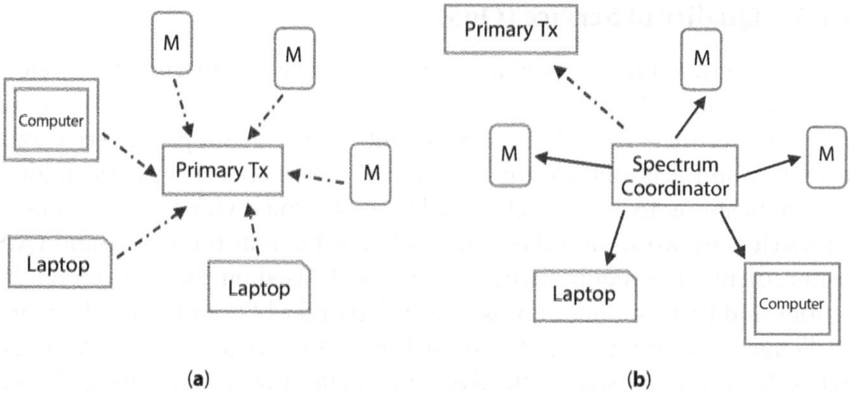

Figure 4.6 (a) Centralized architecture. (b) Distributed architecture.

A. **Adaptive Modulation and Coding (AMC):** mainly this AMC is used to determine the coding and modulation rate based on the number of users and packets. The coding and modulation can be determined by applying the condition of SINR as it provide the improved rate of coding and modulation and this helps to increase the data rate and channel capacity [19].

B. **Spatial Multiplexing:** Using multiple processing, transmitter transmit the information parallel using multiple input stream and parallelism can be applied using multiple antenna and the receiver will receives the information using multiple chain receiver to the destination [20]. The various channels uses multiple path to transmit suing various antenna and it results in decorrelation factor and signal scattering factor. Data rate and channel gain can be determined using spatial multiplexing and it depends on the number of channels usage.

C. **Multi Accessing Technique:** By applying multiple accessing capability, every user in the network has to ensure the effective spectrum utilization and all the resource available in the network has to be effectively shared to all the users [21]. In the resource process, spectrum effectiveness has to determined based on the usage of channel quality information.

4.1.5 Quality of Service (QoS)

There are certain QoS services are applicable while considering the wireless broadband networks. This QoS is termed in the network based on several performance metrics such as, throughput, packet delay, packet error rate and jitter and it is suitable for several real time networking applications. The applications are broadly classified based on voice, video and data based on wireless networks and those network has different traffic rate and QoS requirement. QoS requirement can be varied based on the network applications and there service and users are also differ [22] and [23]. The main challenges is nothing but accommodating all QoS requirement in a single network. Another issue is effectively managing the resource usage based on providing the QoS in the wireless networks. The above issue can be mitigated based on effective process of scheduling the resource among the users and able to balance the QoS requirement in the networks. The scheduling helps to provide the channel usage to the users in equal proportion in the application in a proper time interval [24] and [25]. Another issue is unbalancing handling of simultaneous user in terms of continuous and busty data traffic in the network and it results in the variation of throughput and latency as represented in Figure 4.7.

The QoS and network flow differentiation can be mentioned as the effective signalling mechanism is applied.

A. **Security:**
   Here the wireless broadband technologies such as, WiMAX is having better encryption technique i.e., Advanced Encryption Standard (AES) and it provides security privacy

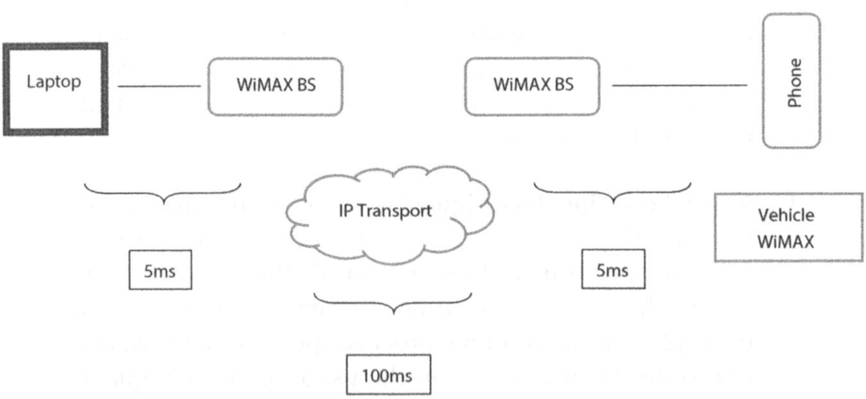

Figure 4.7 QoS in WiMAX architecture.

features with proper key management properties [26]. Then with respect to security and user authentication, EAP architecture is being used and it used certain information's like some of the credentials, of user information, smart card information, etc.

B. **Mobility Support:**
To ensure the support of mobility, seamless handover has to be maintained with proper security with respect to the real-time applications of VoIP. With regards to VoIP, it has a high battery lifetime in order to support a large number of subscribers and this helps in the development of power-saving techniques [27]. The above features are supported in various mobile applications along with various channel estimations on both uplink and downlink.

C. **IP Platform:**
The IP-based architecture is applicable to the network architecture based on a reference as it provides Quality of Service (QoS) to be delivered as a service [28]. There are certain protocols able to provide security and mobility based on transport, session, and QoS management. With respect to WiMAX, IP can be processed, able to communicate with other networks easier and able to deploy the IP easily.

4.1.6 OFDM with WiMAX Technology

With respect to OFDM, there are various physical layer implementations are used for various mobile technologies like mobile WiMAX and WiMAX [29] and [30]. The OFDM parameters are mentioned in Table 4.2.

There are certain other parameter involved in OFDM-PHY and OFDMA-PHY as listed below,

Table 4.2 OFDMA with respect to WiMAX.

S. no.	Technologies	Physical layer
1	Fixed WiMAX	256 FFT-based OFDM
2	Mobile WiMAX	128 to 2,048 FFT-based OFDM

- With respect to fixed WiMAX OFDM-PHY, 256 is the FFT size and it contains 192 subcarriers, 8 pilot subcarriers, and 56 guard band subcarriers. Then the OFDM-PHY contains certain cyclic prefixes with 1/32, 1/16, 1/8, and 1/4.
- For both fixed WiMAX-OFDM-PHY and Mobile WiMAX OFDMA-PHY, the oversample rate can be determined, which depends on certain bandwidth of 256 OFDM for 7/6, 1.75 MHz with the oversampling rate of 28/25 along with 1.5 MHz, 2 MHz, and 2.75 MHz.
- With respect to mobile WiMAX OFDMA-PHY, 512 is the FFT Size and it contains 360 number of subcarriers, 60 pilot subcarriers, and 92 guard band subcarriers. Then OFDMA-PHY contains certain cyclic prefixes with 1/32, 1/16, 1/8, and 1/4.

A. Fixed WiMAX ODFM-PHY:

In these wireless technologies, FFT size is fixed with 256 as it contains a certain subcarrier of 192 with the data carrier communication and 8 pilot subcarrier for the process of estimating channel and synchronization. Then the other carrier channel is used for the purpose of guard band subcarrier. The subcarrier and channel bandwidth have certain variations and have fixed FTT size [31]. As the channel subcarrier increases in space and decreases in the symbol time resulted based on the large bandwidth usage. When the symbol time gets decreased, guard time has to be allocated with a large fraction as it overcomes the spread in the delay time. Robustness of data results in greater delay and effective spectral utilization as it determines the trade-off and makes effective system design.

Based on data robustness, there will be a greater delay in the data communication and it achieves 25 % guard time is used. Based on the usage of multipath channels, it makes reduces the guard time with 3% overhead.

B. Mobile WiMAX OFDMA-PHY:

By considering the Mobile WiMAX technology, with the variation of 128 to 2048 of size for FFT. As the channel subcarrier is spaced with 10.94 kHz when there is an increase in the available bandwidth and also FFT size [32]. The resource unit size will be fixed as the OFDM duration gets updated based on the symbols and also helps in minimizing the scaling value based on the higher value impact. Based on scalability in Mobile WiMAX, the cost of the design gets reduced. For both fixed and mobile WiMAX applications, subcarrier spacing has improved balance so that the delay and Doppler spread are achieved.

4.1.7 Advanced Modulation and Coding Techniques in WiMAX

These Modulation and Coding Schemes (MCS) are added into the system with improved communication as it receives the packets with Bit Error Rate (BER) by the receiver. Then the feedback is sent and transmitted by the receiver by using Channel Quality Indicator (CQI) and it gets reconfigured when new packets get into it [33]. To estimate the channel condition, proper MCS has to be selected for another transmission and it gets initiated by the transmitter. During the MCS selection scheme, performance of each scheme selection based on channel maintenance based on BER. At the receiver end, channel estimation can be done based on MCS scheme and feedback is generated between the transmitter and receiver based on Channel State Information (CSI). Delay between the transmission and channel estimation gets reduced, so that the MCS performance gets improved. During the transmission, which MCS scheme is used those information are sent by the transmitter to the receiver. If the improper modulation coding is used, it will degrade the performance of the receiver [34]. In this MCS scheme, there are some major components has to be discussed,

A. **Channel Estimation:**
   It is a method to calculate the influence of physical medium based on the sequence of input. Based on the transmitter channel estimation, channel impact is determined and those channel are known or not partial, the channel has subcarrier are less spaced [35]. The channel estimation is calculated based on the sequence of input and it will be known to both transmitter and receiver. In order to estimate the channel, training bits and samples received by the receiver.
   - LSE channel estimation
   - LMMSE channel estimation

4.1.8 Conclusion

Based on the component's discussed based on the broadband wireless technologies i.e., WiMAX, the technologies uses MIMO concept to offers effective spectral utilization with higher order precision. It has s salient features of improving mobility based on certain features like seamless connectivity during handover process and uses portable devices, which achieves low power consumption. As WiMAX technology is concern, there are two prominent network topology is considered as mentioned as PMP and Mesh network topology. Then we have discussed the radio channel in terms of LoS and NLoS Connection. Then QoS is discussed with respect to IP mobility, Security and mobility Support. To ensure data robust in WiMAX, encryption process is strong and provide mutual authentication. In order to provide flexibility among the WiMAX technologies, it will support voice, video and multimedia to maintain data traffic among them and ensure QoS.

References

1. I-Shyan Hwang; Jhong-Yue Lee; Chih-Wei Huang; Zen-Der Shyu, Advanced Dynamic Bandwidth Allocation and Scheduling Scheme for the Integrated Architecture of EPON and WiMAX, Tenth International Conference on Mobile Data Management: Systems, Services and Middleware, 2009.
2. Addisu Eshete; Andres Arcia; David Ros; Yuming Jiang, Impact of WiMAX Network Asymmetry on TCP, IEEE Wireless Communications and Networking Conference, 2009.
3. Dongmyoung Kim; Hua Cai; Minsoo Na; Sunghyun Choi, Performance measurement over Mobile WiMAX/IEEE 802.16e network, International Symposium on a World of Wireless, Mobile and Multimedia Networks, 2008.
4. C. Cicconetti; L. Lenzini; E. Mingozzi; C. Eklund, Quality of service support in IEEE 802.16 networks, IEEE Network, Vol. 20, Issue. 2, pp. 50-55, 2006.
5. L. Berlemann; C. Hoymann; G.R. Hiertz; S. Mangold, Coexistence and Interworking of IEEE 802.16 and IEEE 802.11(e), IEEE 63rd Vehicular Technology Conference, 2006.
6. Qinghua Li; Xintian Eddie Lin; Jianzhong Zhang; Wonil Roh, Advancement of MIMO technology in WiMAX: from IEEE 802.16d/e/j to 802.16m, IEEE Communications Magazine, Vol. 47, Issue. 6, pp. 100-107, 2009.
7. Zeeshan Ahmed, Salima Hamma & Zafar Nasir, An optimal bandwidth allocation algorithm for improving QoS in WiMAX, Multimedia Tools and Applications, Vol. 78, pp. 25937–25976, 2019.
8. M. Deva Priya, M. Sangeetha, A. Christy Jeba Malar, E. Dhivyaprabha, N. Kiruthiga & PL. Rajarajeswari, Fair Adaptive Cross-Layer Resource Allocation Scheme for IEEE 802.16 Broadband Wireless Networks, Wireless Personal Communications volume, Vol. 117, pp. 2645–2666, 2021.
9. Mahboubeh Afzali, Kamalrulnizam AbuBakar & Jaime Lloret, Wireless Personal Communications, Vol. 107, pp. 849–867, 2019.

10. Sahar Abdul Aziz Al-Talib and Noor Raad Saadallah, Data Path Management Mechanism to Mesh Base Stations in WiMAX, American Journal of Engineering and Applied Sciences, Vol. 12, Issue. 1, pp. 67-77, 2019.
11. Angeliki Alexiou, Wireless World 2020: Radio Interface Challenges and Technology Enablers, IEEE Vehicular Technology Magazine, Vol. 9, Issue. 1, pp. 46-53, 2014.
12. Cheng-Xiang Wang; Fourat Haider; Xiqi Gao; Xiao-Hu You; Yang Yang; Dongfeng Yuan; Hadi M. Aggoune; Harald Haas; Simon Fletcher; Erol Hepsaydir, Cellular architecture and key technologies for 5G wireless communication networks, IEEE Communications Magazine, Vol. 52, Issue. 2, pp. 122-130, 2014.
13. Haider Rasheed Abdulshaheed, Zeyad Taha Yaseen, Adil M. Salman and Israa Al_Barazanchi, A survey on the use of WiMAX and Wi-Fi on Vehicular Ad-Hoc Networks (VANETs), IOP Conference Series: Materials Science and Engineering, Vol. 870, 2020.
14. Jayant Baliga; Robert Ayre; Kerry Hinton; Rodney S. Tucker, Energy consumption in wired and wireless access networks, IEEE Communications Magazine, Vol. 49, Issue. 6, pp. 70-77, 2011.
15. Shahad Nafea and Ekhlas Kadum Hamza, Path loss Optimization in WIMAX Network using Genetic Algorithm, Iraqi Journal of Computers, Communications, Control & Systems Engineering, Vol. 20, No. 1, pp. 24-30, 2020.
16. Agbotiname Lucky Imoize and Tolulope Emmanuel Ogunfuwa, Propagation measurements of a 4G LTE network in Lagoon environment, Nigerian Journal of Technological Development, Vol. 16, No. 1, 2019.
17. Eugen Harinda; Salaheddin Hosseinzadeh; Hadi Larijani; Ryan M. Gibson, Comparative Performance Analysis of Empirical Propagation Models for LoRaWAN 868MHz in an Urban Scenario, IEEE 5th World Forum on Internet of Things (WF-IoT), 2019.
18. Rajanandhini C and Babu S.P.K., Analysis of UWB MB-OFDM Communication System through IEEE802.15.4a Channel Models using Matlab Simulink, International Conference on Recent Trends in Computing, Communication & Networking Technologies (ICRTCCNT), 2019.
19. Lavish Kansal, Vishal Sharma & Jagjit Singh, Model of adaptive WiMAX network incorporating diverse transforms, Wireless Networks, Vol. 25, pp. 3497–3511, 2019.
20. Behnam Kamali, Adaptive Modulation-Coding Scheme for Link Adaptation in Aeronautical Mobile Airport Communications Systems (Aeromacs), Integrated Communications, Navigation and Surveillance Conference (ICNS), 2019.
21. Mohamed Asma , Zekry Abdel Halim , Ibrahim Reem, FPGA Implementation of Sphere Detector for Spatial Multiplexing MIMO System, International Journal of Electronics and Telecommunications, Vol. 65, No. 2, pp. 245-252, 2019.

22. M.M.Eissa, Medhat H.A. Awadalla, Centralized protection scheme for smart grid integrated with multiple renewable resources using Internet of Energy, Global Transitions, Vol. 1, pp. 50-60, 2019.
23. Mustafa Qays Hatem, Ahmed Aizaldeen Abdullah, Quality of Service of VoIP over WiMAX, Journal of Southwest Jiaotong University, Vol. 54, No. 4, 2019.
24. Deepa Naik, Santosh Dora, Nikita, Tanmay De, Normalized Uplink Bandwidth Scheduling Algorithm for WiMAX Networks, Advances in Computer, Communication and Control, pp 311-325, 2019.
25. Zeeshan Ahmed, Salima Hamma & Zafar Nasir, An optimal bandwidth allocation algorithm for improving QoS in WiMAX, Multimedia Tools and Applications, Vol. 78, pp. 25937–25976, 2019.
26. Sankaranarayanan, P; Gayathri, R, Distributed Mobility Management Scheme for All IP Based Next Generation Heterogeneous Networks, Journal of Computational and Theoretical Nanoscience, Vol. 16, No. 4, pp. 1332-1334, 2019.
27. Ahmed Shaban Samrah, Abeer Twakol Khalil and Hegazi M. Ibrahim, Improving quality of service for internet protocol television and voice over internet protocol over long-term evolution networks, International Journal of Communication Networks and Distributed Systems, Vol. 22, No. 4, pp. 409-446, 2019.
28. V. Dhilip Kumar, G. R. Kanagachidambaresan, Phidahunlang Chyne & D. Kandar, Extended Communication Range for Autonomous Vehicles using Hybrid DSRC/WiMAX Technology, Wireless Personal Communications, 2021.
29. Varshini Rajesh, A. R. Abdul Rajak, Cancellation Techniques for Co-channel Interference in MIMO-OFDM Systems and Evaluating Their Performance, Emerging Science Journal, Vol. 5, No. 6, 2021.
30. Lavish Kansal, Gurjot Singh Gaba, Ashutosh Sharma, Gaurav Dhiman, Mohammed Baz, and Mehedi Masud, Performance Analysis of WOFDM-WiMAX Integrating Diverse Wavelets for 5G Applications, Wireless Communications and Mobile Computing, 2021.
31. Debdatta Kandar,V. Dhilip Kumar,Sukumar Nandi, Smart inter-operable vehicular communication using hybrid IEEE 802.11p, IEEE 802.16d/e technology, International Journal of Communication systems, Vol. 34, Issue. 11, pp. e4847, 2021.
32. Siwatch Sonia, Saini Garima, Performance evaluation of a WIMAX system using coding technique, International Journal of Managment, IT and Engineering, Vol. 2, Issue. 11, pp. 555-563, 2012.
33. Aymen Mohammed Khodayer Al-Dulaimi; Mohammed Khodayer Hassan Al-Dulaimi; Omer Mohammed Khodayer Al-Dulaimi, Construction and Analysis of Dynamic Distribution for Resource Blocks of Real-Time and Data Elastic Traffic in IMS/LTE networks, International Congress of Advanced Technology and Engineering (ICOTEN), 2021.

34. João Martins, Filipe Conceição, Marco Gomes, Vitor Silva and Rui Dinis, Joint Channel Estimation and Synchronization Techniques for Time-Interleaved Block-Windowed Burst OFDM, Applied Sciences, Vol. 11, Issue. 10, 2021.
35. XiFang, Zhufeng Suo, Lei Zhang, Qingchuan Zhang, FanZhang, Combined phase offset channel estimation method for optical OFDM/OQAM, Optical Fiber Technology, Vol. 61, 2021.

5

Network Routing and Its Real-Time Practice in Broadband Wireless Networks

R. Prabha[1]*, G.A. Senthil[2†], S. K. B. Sangeetha[3], S.U. Suganthi[4] and D. Roopa[5]

[1]*Department of Electronics and Communication Engineering, Sri Sairam Institute of Technology, Chennai, India*
[2]*Department of Information Technology, Agni College of Technology, Chennai, India*
[3]*Department of Computer Science and Engineering, SRM Institute of Science &Technology, Chennai, India*
[4]*Department of Artificial Intelligence and Data Science, Sri Sairam Institute of Technology, Chennai, India*
[5]*Department of Computer Science and Engineering, Sri Sairam Institute of Technology, Chennai, India*

Abstract

Wireless technologies that enable people to link their home networks together as well as to the Internet via wireless community networks are in high demand. Despite the fact that many network architectures for broadband wireless networking have been studied and implemented in practice due to emerging technology and locales, Broadband Wireless Networking (BWN) has unquestionably been the most common architecture that has demonstrated strong dominance in various roles. Indeed, allowing such broadband networking and creating a group network has numerous advantages. This would boost the rural and isolated communities' quality of life, digital interaction, and collaborations. The design of efficient communication is a difficult problem for the next generation BWNs' performance in managing real-time and QoS-sensitive applications, as well as pleasing both service providers and consumers. Unfortunately, today's cutting-edge routing standards in BWNs are not ideally suited to tackle this task, as these standards are fundamentally complex and suffer from innate problems with regard to efficient communication-based implementations, according to the literature. As a result, the aim of this chapter is to provide BWNs with a straightforward roadmap of theoretical context so that they can manage various efficient routing mechanisms.

**Corresponding author*: r.praba05@gmail.com
†*Corresponding author*: senthilga@gmail.com

A. Suresh, J. Ramkumar, M. Baskar and Ali Kashif Bashir (eds.) Resource Management in Advanced Wireless Networks, (99–112) © 2023 Scrivener Publishing LLC

Keywords: Broadband wireless networking, network routing, broadband, quality of service, WiMax

5.1 Introduction

Remote broadband terms ought not be related with the normal word broadband availability or BISDN (Broadband Integrated Services Digital Network), which applies to various correspondence organizations (fiber or optical) utilized by ISPs and NSPs to arrive at Internet spine information speed of in excess of 155 Mbps. BISDN is the wire and connection that goes through dividers, under floors, from utility post to utility shaft, and under feet on a city street in layman's terms. BISDN is a perspective, a lot of developments, and a lot of norms for joining progressed correspondence organizations into a fiber optic and radio-based broadband network [1]. Data correspondence organization for fast data sent in wide detonates, the Fiber Distributed-Data Interface (FDDI), and the Synchronous Optical Network are exceptionally significant for BISDN (SONET). BISDN can impart data at speeds going from 2 Mbps to a lot more noteworthy. Remote broadband, on either hand, is a distant association structure that handles irrefutably the last issue by allowing us to interface disengaged customer premises to an ISP or carrier's establishment network without leasing customary T-1 and more significant copper or fiber stations from the local telecom provider. Fixed remote access that can be used by associations, endeavors, families, and remote workers who move beginning with one unequivocal spot then onto the following is implied as remote broadband. It doesn't resolve the issues of convenient customers making the rounds in its present structure [2]. Remote broadband is a utilitarian extension of the highlight point, remote LAN combination idea for conveying high velocity, high-limit pipes for voice, mixed media, and Internet network administrations. Albeit remote broadband is essentially used to interface LANs to the Network in straightforward applications, further developed executions permit you to associate various administrations (information, voice, and video) over a similar link. The last requires the establishment of multiplexing offices at the client's area or in a focal hub. Wireless Broadband detours actual phone networks as far as organization, and it is similarly as attainable in country regions all things considered in metropolitan regions. Seller arrangements that keep away from costly establishment, overhauling, and redesigns mean skirting 120 years of broadband improvement for geologies that haven't yet innovatively developed to fiber and cabling frameworks. Liberation in certain spaces has worked on the permitting system for Wireless Service Providers (WSPs) [3].

Remote broadband is faster to market, and clients are presented in stages rather than all at once, avoiding the costly upgrades that must be completed before wired subscribers can communicate. Wireless broadband is intended for fixed wireless connections; it does not meet mobility requirements, which are currently addressed by only 2.5 G and 3G networks. Broadband wireless radios may potentially be miniaturized and integrated into portable devices in the future. They would then be able to supplement 3G in mobile apps. However, none of the vendors have any getting experience in this field, so it's just a physics and circuitry possibility. Wireless broadband, of course, is intended to meet the needs of suburban Internet access by bypassing local telcos. Multiple research projects are underway to improve WOBAN, such as the management system and energy conservation [4]. As a result, the WOBAN architecture acts as a foundation for future discussion of the aforementioned topics. WOBAN is also known as FiWi BAN by some researchers (Fiber Wireless Broadband Access Networks). We use the same language as the source to ensure continuity between this survey and the references. As a result, when discussing related works, the word FiWi BAN can appear.

The architecture of BWA networks in general covers a wide range of topics. First and foremost, the access network's bandwidth must be increased. Optimal relay station positioning, bandwidth allocation scheme, and direct communication mode between subscriber stations have all been proposed. Other related research focuses on large-scale BWA network management systems and efficiency comparisons with multiplex strategies for 4G [5]. Another important consideration when installing BWA networks is energy and cost management. Reducing energy usage not only saves money on electricity bills, but it also helps to minimize CO2 emissions and is environmentally friendly. Several studies have been conducted on this subject, including an architectural study of energy efficiency, a mixed capacity access proposal, a study for minimizing emissions in long reach access, and a green WOBAN design. The enhancement of survivability and real application situations that can benefit from BWN networks are also important topics in BWN [6].

With this detailed introduction about BWN, Section 5.2 describes the outline of BWN, Section 5.3 discusses various routing mechanisms in BWN, Section 5.4 briefs about security issues in BWN followed by conclusion in Section 5.5.

5.2 Outline of Broadband Wireless Networking

5.2.1 Type of Broadband Wireless Networks

5.2.1.1 Fixed Networks

Rapid remote organizations that connect to fixed areas and are intended to help itinerant clients are known as fixed broadband remote innovations. Fixed remote innovations incorporate Wireless Fidelity (Wi-Fi) and Worldwide Interoperability for Microwave Access (WiMax). IEEE 802.11 and IEEE 802.16 guidelines control these two advancements, respectively [7, 8].

Wi-Fi was the first high-speed fixed wireless broadband technology to get into the industry. The first WLAN (802.11) was implemented in 1997 and could accommodate 2 Mbps, while IEEE accepted 802.11b in 1999. There are currently several wireless devices based on Wi-Fi technology, including the IEEE 802.11a, b, and g standards, as well as an 802.11n standard that has yet to be defined. It's difficult to say what effect 802.11n would have when it actually hits the market, but it's expected to offer up to 600 Mbps. In a 2006 survey, In-Stat predicted that more than half of all chipsets shipped in 2008 will be focused on 802.11n. Wi-Fi has seen widespread use as a high-speed wireless technology, most notably in hotspots around the world.

5.2.1.2 WiMAX

WiMax (Worldwide Interoperability for Microwave Access) is another proper broadband remote innovation that will give last-mile broadband inclusion across a more extensive topographical region than Wi-Fi. It should involve a space of one to six miles in measurement. WiMax inclusion scopes of this size are planned to give fixed and traveling remote broadband access without the requirement for a line-of-site (LOS) with a base station. WiMax can likewise have greater portability, higher-speed information applications, more prominent inclusion, and throughput than Wi-Fi. The execution of WiMax can give various advantages. Regardless, it considers higher throughput rates, quicker information speeds, and a more extensive working reach. Thus, the innovation is great for sending in troublesome landscape or in regions with deficient wired framework.

WiMax additionally upholds and effectively incorporates with other wired and remote technology. The greatest inconvenience of WiMax execution is the utilization of restrictive gear. To give ideal usefulness, WiMax hardware should have the option to productively use power. The yield power utilized by WiMax is controlled by a running technique that indicates the appropriate planning offset and force settings. Therefore, endorser station's transmissions are relied upon to show up at the base station simultaneously and with a similar force level. At the point when WiMax is utilized outside, it can encounter delays in non-view conditions, which can prompt intersymbol interference.

5.2.1.3 The Broadband Mobile Wireless Networks

The first (1G) associations, which depended on Frequency Division Multiple Access (FDMA) and were by and large used for voice correspondence, begun the advancement of convenient help. Second-age (2G) networks have for the most part supplanted the 1G organization, which is for the most part utilized for discourse applications. At a low speed, these 2G organizations conveyed circuit-exchanged information transmission administrations. The scramble to create and consolidate advanced frameworks brought about a plenty of divergent and contradictory specifications. 2.5G is an improvement of 2G innovation that takes into account more information transmission capacity on 2G organizations. Hence, progresses like General Packet Radio Service (GPRS) and Enhanced Data Rates for Global Evolution (EDGE) were made (EDGE). The point of third-age (3G) networks was to give a typical general norm to rapid information and top-notch voice administrations. The point was for all clients all throughout the planet to utilize a typical norm, taking into account genuine worldwide meandering. Soon, 4G is a cutting-edge wave of broadband that will enhance and supplant 3G organizations. The principal provisions of 4G foundations are getting to data anyplace, anyplace, with a consistent connection to a wide assortment of data and benefits, and getting a lot of data, information, pictures, video, etc. Future 4G frameworks would be comprised of an assortment of organizations that utilization IP (Internet convention) as a standard convention, giving clients unlimited oversight about the applications and conditions they use.

In view of ebb and flow portable availability designs, 4G would highlight expanded limit, higher information rates, and smoother and quicker handoff, with an accentuation on giving consistent inclusion through an assortment of remote frameworks and organizations. The main idea is to use new technology to integrate 4G features with all current mobile technologies. Some of the key aspects of 4G networks that consumers are interested in are application adaptability and high dynamism. These features imply that services can be provided and made available according to the preferences of individual customers, while still supporting traffic, air interfaces, radio environments, and service quality. The right and efficient link to network applications can be done in a variety of ways and at different levels [9].

5.2.2 BWN Network Structure

The BWN network is principally utilized as a highlight multipoint geography, with a cell conveyance of base stations that are totally associated with center organizations and in correspondence with fixed remote supporter stations. The endorser stations as a rule include housetop mounted receiving wires and radio units connected to indoor organization interface units, while the base stations are usually found on a pole. Multiple subnets in a star topology usually make up a completed BWN structure. In terms of structure, it resembles the historic slotted Aloha network proposed in the 1970s. The downstream of the vehicle system, from the central base station to the endorser stations, is sent, while the upstream, from the supporter stations to the central base station, utilizes time division various access (TDMA). The organization's fundamental parts are portrayed in Figure 5.1.

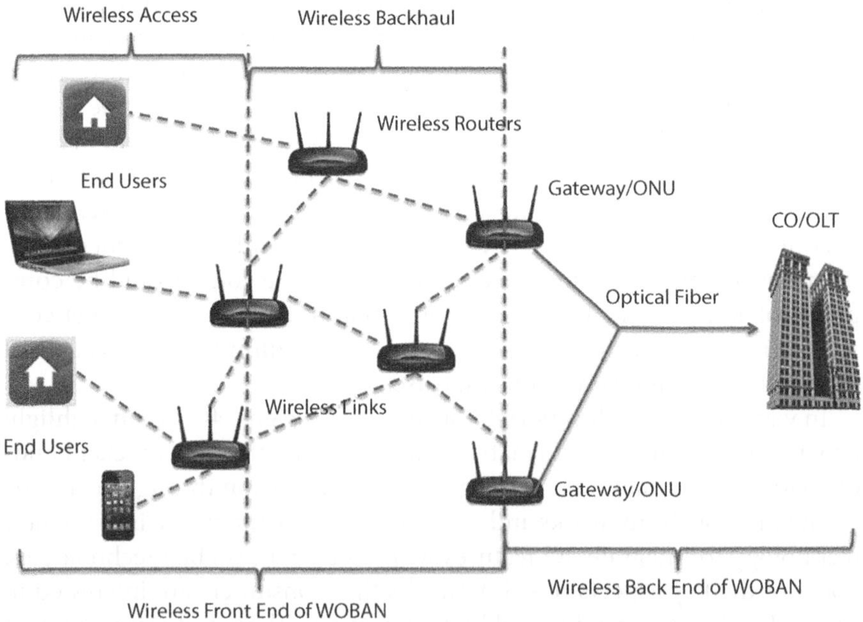

Figure 5.1 Structure of WBN.

Thousands of customer premises equipment (CPE) can be found on the rooftops of various structures. Optical fibre can link the central base station with other server applications to the Internet backbone. The base station (BS) is a switching system that communicates directly with the subscriber station and is also referred to as a network bridge or router. The modem will serve as a data port on the subscriber station (SS), resuming data transmission over cable. The physical and data link layers of a BWA network are the two layers that differentiate it from other networks, according to the OSI model. They define the methods and specifications for communication between the Subscriber Station (SS) modem and the Base Station (BS). A BWA system's network protocol closely follows the DOCSIS (Data Over Cable Service Interface Specifications) format, with some modifications for wireless use [10].

5.2.3 Wireless Broadband Applications

Broadband communication refers to the transmission of digital audio (voice), data, and/or video at speeds faster than those of wideband communication (above 1 Mbps). Broadband connections enable the delivery of multiple services such as voice, data, and video over a single network [11].

5.2.3.1 Digital Telephone

Digital telephony is a communication device that represents and transmits analogue signals using digital data. Audio signals (acoustic sounds) or complex modem signals (other types of information) are examples of analogue signals. WiMax systems can use IP Telephony to provide telecommunications services (voice over Internet protocol VoIP). These IP networks use IP protocol to initiate, process, and receive voice or digital telephone communications. WiMax systems may use analogue telephone adapters (ATAs) or IP telephones to provide wireless telephone service. IP signals are converted into regular telephone (dial tone) formats by ATAs.

5.2.3.2 Broadband Data Connections

Broadband data networks are those that transmit digital data signals at speeds of 1 megabit per second or higher for consumer connections and 45 megabits per second or higher for LANs, MANs, and WANs. WiMax service providers can be referred to as wireless Internet service providers because they provide broadband data services that can link to the Internet (WISP).

5.2.3.3 *Digital Television*

The method or device that uses digital transmission to relay video images is known as digital television. Channels for digital video and audio are split into the digital transmission. Typically, these digital channels are compact.

5.2.4 Promising Approaches Beyond BWN

To meet future capacity demands, radio access strategies will need to be significantly improved. There is no easy alternative; instead, the solution would be a mixture of multiple options, depending on the time and place. In networking, radio access, and modem integration, there are some promising strategies to take, which are briefly outlined below.

- Increases in capacity of 500-1000 times are needed.
- There is a scarcity of spectrum.
- Demands for energy conservation small cells are being used on a large scale. Super-broadband optical fibre is now available.
- Spectrum use that is adaptable.

Every year, the market for broadband wireless communications grows dramatically across the world. The ever-increasing number of users who subscribe to broadband packages is a major factor in this growth. The movement toward flat-rate contracts has intensified this. Furthermore, new devices with strong multimedia capabilities, such as mobile phones and tablets, are entering the market, generating new demands for broadband wireless connectivity. Finally, new data services and apps are becoming available, which are critical success factors for mobile broadband. As both of these variables are combined, the data traffic in the wireless access system grows exponentially.

Over the next decade, this pattern is projected to continue in a similar manner. According to recent studies and extrapolations based on historical trends, overall traffic will rise by a factor of 50 to 10000 in the next period. Aside from the total traffic, the realistic throughput per user must be greatly improved. According to a rough estimate, the average and peak data rates would rise by at least ten times. Furthermore, important design standards that must be met more effectively than in today's systems include user fairness across the entire coverage area, latency to minimize response time, and better support for a variety of Quality of Service (QoS) requirements originating from various networks [12].

5.3 Routing Mechanisms

The directing convention used to track down the best course to the base stations is presently being investigated as one of the vital spaces of cross section organizations (or passageways). Clients who utilize this innovation can exploit their administrations all the more adequately and impart all the more proficiently, just as move their information stream through the remote correspondence climate. Directing is an assistance wherein the switch analyzes the entirety of the accessible ways for sending bundles to their objective and picks the right one. The rule of organization execution improvement is carried out by developing a steering tree choice that is characterized by topological properties that are independent of the organization's arrangement. The manner in which the tree is constructed and the hubs are masterminded takes into account a superior appropriation of hubs, which prompts better directing and collection. The likenesses between the tree's mathematical boundaries and the organization's proficiency should be assessed, and those with the best relationships ought to empower the best trees to be made, taking into consideration some directing and geography improvement [16, 17].

There is right now a scope of steering conventions accessible, each with its own arrangement of benefits and detriments when applied to work organizations. There is no single directing convention that can be supposed to be awesome among such countless choices. The clarification for this is that they each have their own arrangement of attributes, and there is no single convention that is viewed as ideal in all situations. Every convention has a distinctive component that makes it proper for a particular application [18].

Adaptive routing algorithms can in turn be classified in two ways:

5.3.1 Distance Vector (DV)

This became known as Routing Information Protocol (RIP) or Distributed Bellman-Ford Protocol (DBF) due to its application to bundle directing on the Internet (DBF). This calculation works by permitting every switch to keep a table (for example a vector) of the most brief distances to each known objective and figuring out which line to take to arrive. Steering is characterized as a metric unit in a distance vector that addresses the expense of a way between network hubs. The actual distance between hubs, the quantity of bounces (jumps), the transmission inactivity, hub clog, and different factors could all be remembered for this metric unit.

5.3.2 Link State (LS)

This powerful calculation was made to tackle the issue of distance vector steering since it depended on the quantity of jumps to the objective, despite the fact that a bundle could arrive at its objective by voyaging a brief distance, for example with few bounces. Be that as it may, the association data transfer capacity can be limited, bringing about a more extended deferral. As a result, the connection state has emerged in order to find efficient routes, regardless of the total number of hops in the network is situated.

Routing protocols for ad-hoc networks are divided into three categories: constructive, reactive, and hybrid. When a data packet needs to be sent, the proactive form requires us to maintain the route network for all possible destinations. The nodes in reactive protocols discover the destinations on request. The hybrid protocols are those in which there is only one group of nodes that updates information on potential destinations on a regular basis. In BWN, QoS routing is a critical parameter for ensuring guaranteed QoS. In BWN, this problem has been thoroughly investigated. The point of QoS directing in these organizations is twofold: to track down the most ideal course for every approaching association within the sight of hidden connection, and to adjust the heap to augment network use.

Traditional routing protocols use the number of hops as their most common criterion. Notwithstanding, obviously these conventions are deficient for media applications that need QoS affirmations, for example, VoIP and video conferencing. Steering conventions with QoS should track down the most brief, yet additionally the best course that fulfills the start to finish QoS boundaries, independent of the measure of bounces or how the correspondence conventions should track down the best courses across various jumps. It is important that the most recent conventions and directing calculations consider boundaries and different measurements, for example, power utilization, the nearness of the spine network yield, and, specifically, the quality over amount association for clients and the nature of remote interchanges, while representing weakening, signal quality, and disturbance [13, 14].

5.4 Security Issues and Mechanisms in BWN

5.4.1 DoS Attack

Quite possibly the most major issues with a wide range of remote network, particularly broadband remote organizations, is forswearing of administration (DoS). A DoS break happens when enrolled clients are not furnished with a mentioned administration inside a characterized greatest holding up period. The most problematic and damaging assault can be directed on any layer of a broadband remote organization. DoS assaults expect to upset accessibility by impeding contact between network gadgets or keeping a solitary gadget from sending or getting traffic; accessibility guarantees that supported clients can get to information, administrations, and organization assets from any area whenever.

A DoS attack on the actual layer might be dispatched by utilizing a radio sticking framework or a wellspring of solid commotion to meddle with the actual channels, conceivably risking administration accessibility. Nonetheless, this kind of assault is remarkable on the grounds that it requires the utilization of specific equipment, and sticking assaults can be identified utilizing radio analyzers. It may cause major issues during the sharing of confidential information or during combat. None of the mechanisms are up to the task of dealing with a jamming assault on these BWN.

5.4.2 Distributed Flooding DoS

A distributed flooding DoS attack poses a significant threat to all wireless broadband networks, as it has the potential to bring an entire network down or consume a significant amount of network bandwidth. This type of attack begins by infiltrating a wireless network with a large number of innocent nodes known as Zombies, which are programmed by highly qualified programmers. These zombies submit data to pre-selected attack targets, clogging the network in the process. DDoS attacks are almost impossible to avoid in most situations, and they have the potential to flood and overflow networks.

5.4.3 Rogue and Selfish Backbone Devices

By breaching the core network equipment, the intruder will severely damage broadband wireless networks. A rogue BS, as defined by IEEE 802.16, is an intruder station that is used to confuse the network's mobile stations by appearing and acting like a legitimate BS. Sniffers are used by attackers to hack mesh routers or access points. A sniffer is a programmed that analyses network traffic passively as part of a passive traffic analysis attack. The BS is taken advantage of in IEEE 802.16 by reconstructing a PC with the equipment address of one more substantial gadget, which can be identified by capturing IEEE 802.1 administration messages with sniffers. A similar strategy can be utilized to think twice about switches and passages by utilizing the equipment address of another organization PC.

5.4.4 Authorization Flooding on Backbone Devices

Test demand outlines are utilized by WBN and IEEE 802.11 hubs to find a remote organization, and in the event that one exists, the AP reacts with a Probe reaction outline. Customers pick the AP that conveys the best message to them. The interloper might parody a surge of test demand outlines, representing an enormous number of hubs searching for a remote organization, making the AP or remote lattice switch be seriously overloaded. If the heap arrives at the limit esteem, the AP or remote cross section switch will quit reacting, possibly bringing about help blackout. Customer stations use testaments to validate and enlist with the BS in IEEE 802.16. The customer station's capacity to present an enormous number of enlistment solicitations to the BS which causes a DoS.

5.4.5 Node Deprivation Attack

The aggressors focus on a solitary hub in a hub hardship assault and keep it from taking part in normal organization tasks. In WBN and IEEE 802.11, hubs should initially confirm with the lattice switch or passageway, and afterward de-verify in the event that they at this point don't have any desire to utilize the organization administrations. The attackers will mimic.The attacker will impersonate the target node and spoof the de-authentication message to prevent it from accessing network resources [15].

5.5 Conclusion

Strong research efforts on emerging evolutionary and creative radio access network solutions are essential with the dramatic traffic increase expected for 2025 and beyond. On a global scale, these initiatives will require coordinated activities between industry and academia. Broadband wireless networking is establishing itself as a legal local access network for high-quality digital data, video, and voice services. It is hoped that through this chapter, which provides an introductory explanation of the BWN network, provides a complete outline of background, routing protocols and security issues. We may consider various issues relating to the security aspect of broadband technology based on the above review. There are many perspectives to consider when considering the security of wireless technologies. Safety encompasses a wide range of authentication, access control, and encryption technologies. While these are important and huge structure blocks for overall protection, they are not the subject of this section and will be considered in future work.

References

1. Craig, K. (2021). Propagation studies for enhanced broadband wireless access.
2. Ibikunle, Francis. (2013). Broadband Wireless Access Deployment Approach to Rural Communities.
3. Guan, Mingxiang & Wang, Le. (2014). Comparison of Broadband Wireless Access Technology for HAPS Communication. Sensors and Transducers. 166. 122-127.
4. Hwang, Kuowei & Vemuri, Rao. (2021). Broadband Wireless Access (BWA) Networks: A Tutorial.
5. Kaydenko, M.M. (2013). Adaptive modulation and coding in broadband wireless access systems. 275-276.
6. Lazov, Igor. (2015). Entropy Analysis of Broadband Wireless Access Systems. IEEE Systems Journal. PP. 1-8. 10.1109/JSYST.2015.2456941.
7. Lazov, Igor. (2017). An Uncertainty Quantification methodology for Broadband Wireless Access systems. Pervasive and Mobile Computing. 42. 10.1016/j.pmcj.2017.10.002.
8. Lin, Hui-Tang & Lin, Ying-You & Kang, Hung-Jung. (2013). Adaptive Network Coding for Broadband Wireless Access Networks. IEEE Transactions on Parallel and Distributed Systems. 24. 4-18. 10.1109/TPDS.2012.101.
9. Liu, Yaqiang & O'Leary, Paul. (2013). Broadband wireless access in an energy efficient environment. 1-18. 10.1049/ic.2013.0097.
10. Nasser, Nidal & Miller, Reid & Esmailpour, Amir & Taha, Abd-Elhamid & Bejaoui, Tarek. (2014). Optimized bandwidth allocation in broadband wireless access networks. Wireless Communications and Mobile Computing. 15. 10.1002/wcm.2479.
11. Nirmalathas, Ampalavanapillai. (2012). Greening the Broadband Wireless Access Networks. 10.1364/ANIC.2012.AW3A.1.
12. Okello, Dorothy & Wasswa, Wilson & Mukasa, Peter & Sebbaale, Derrick & Kagarura, Geoffrey. (2015). eInfrastructure: Next generation wireless broadband networks for Uganda 2020. 10.1109/ISTAFRICA.2015.7190548.
13. Osseiran, Afif & Braun, Volker & Hidekazu, Taoka & Marsch, Patrick & Schotten, Hans & Tullberg, Hugo & Uusitalo, Mikko & Schellmann, Malte. (2013). The Foundation of the Mobile and Wireless Communications System for 2020 and Beyond: Challenges, Enablers and Technology Solutions. 1-5. 10.1109/VTCSpring.2013.6692781.
14. Wake, David & Beacham, Keith. (2021). Passive electroabsorption transceivers for broadband wireless access.

15. Yonis, Aws. (2014). Broadband Wireless Access Networks for 4G.
16. G. A. Senthil, R. Prabha, D. Roopa, D. V. Babu and S. Suganthi, "Improved Cluster Head Selection for Data Aggregation in Sensor Networks," *2021* 7th International Conference on Advanced Computing and Communication Systems (ICACCS), 2021, pp. 1356-1362, doi: 10.1109/ICACCS51430.2021.9442048.
17. Zander, Jens & Mähönen, Petri. (2013). Riding the Data Tsunami in the Cloud: Myths and Challenges in Future Wireless Access. IEEE Communications Magazine. 51. 145-151. 10.1109/MCOM.2013.6476879.
18. S. U. Suganthi, G. Valarmathi, V. Subashini, R. Janaki, R. Prabha, "Coal mine safety system for mining workers using LORA and WUSN", Materials Today: Proceedings, Volume 46, Part 9, 2021, Pages 3803-3808, ISSN 2214-7853, https://doi.org/10.1016/j.matpr.2021.02.037.

6

Routing Mechanism in Broadband Wireless Network

T. Rajendran[1], B. Yamini[2*], K.E. Narayana[3] and K. Pradeep Mohan Kumar[4]

[1]*Department of Computer Science and Engineering, Rajalakshmi Institute of Technology, Chennai, Tamil Nadu, India*
[2]*Department of Networking and Communications, Faculty of Engineering and Technology, SRM Institute of Science and Technology, Kattankulathur, Chengalpattu District, Tamil Nadu, India*
[3]*Department of Computer Science and Engineering, Rajalakshmi Engineering College, Chennai, Tamil Nadu, India*
[4]*Department of Computing Technologies, Faculty of Engineering and Technology, SRM Institute of Science and Technology, Kattankulathur, Chengalpattu District, Tamil Nadu, India*

Abstract

The evolution of broadband services through mobile networks will be biggest boon to all remote village users and the industries which operating far away from major cities. Broadband Wireless Networks (BWN) a networking technology designed to impart high speed Internet and data service through wireless networks. Routing protocols are essential to finding a low-cost route path between the source and targeted nodes through several hops and also adhere with the two important characteristics, robustness and scalability. When the revolution spins around the mobile broadband technologies, there would be huge demand in the requirements for advanced routing mechanisms. There are several challenges associated with the routing protocols that could be hand-in-hand with the massive growth in the technological advancements with 4G and 5G mobile networks. The scale-up in the nodes and hops count is drastically increased. To support this sudden growth, the associated resources and their impacted protocols also need to enhance. Routing protocols customized based on the recent advancements in the delivery of Quality of Services (QoS) and their bandwidth demands. Convergence

**Corresponding author*: yamini.subagani@gmail.com

A. Suresh, J. Ramkumar, M. Baskar and Ali Kashif Bashir (eds.) Resource Management in Advanced Wireless Networks, (113–140) © 2023 Scrivener Publishing LLC

on broadband networks and the service delivery model seems to be more challenge to converge among routing paths across diversified routing models.

The proposed chapter discussed on the following topics:

1. **Classification of BWN Routing Protocols**
2. **Routing mechanism in WiMax**
3. **Routing mechanism in Mobile Networks**
4. **Service Specific Routing Protocols**
5. **Novel approaches and algorithms on broadband routing**

Keywords: Broadband wireless, routing protocols, WiMAX, mobile broadband, convergence communication

6.1 Classification of BWN Routing Protocols

6.1.1 Routing

Routing specifies the route through which data transfer flows within or between networks. This routing may be of Unicast, Broadcast and Multicast routing as given in Figure 6.1. In Unicast Routing, messages will be sent specifically to a particular node and its counterpart is the Broadband Routing where in the messages will be delivered to all available nodes in the network whereas Multicast Routing is intended to send messages to some set of specific mentioned nodes [1].

6.1.2 Routing-Protocols

Routing-Protocols are distinct procedures collection that routers follow for establishing communication by exchanging the routing information

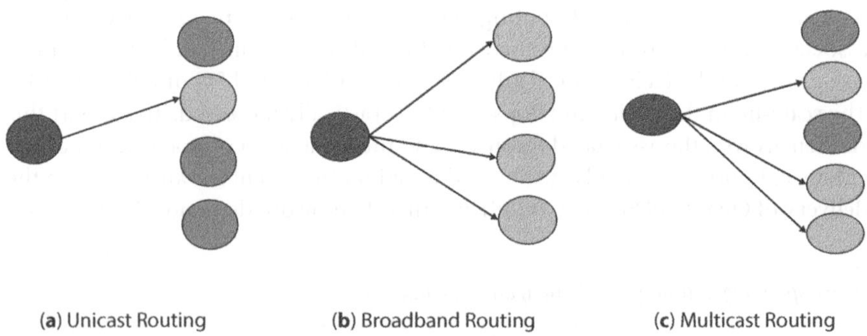

Figure 6.1 Types of routing mechanisms.

and dynamically updating the routing table for specific route identification between the source & destination. It helps in recognizing all live available routes between source & destination.

6.1.2.1 Types of Routing Protocols for Broadband Wireless Network

Based on the topological structure of the network, Broadband Wireless Network Routing protocols are categorized as Static, Default and Dynamic Routing Protocols [2] as mentioned in the Figure 6.2.

Static Routing Protocol in Broadband Wireless Network

Static Routing Protocols can be converged with Dynamic Routing Protocols. These protocols facilitates to reach a specific network by manually updating the routing table. Static Routing Protocol provides enhanced security for data transmission over the network because of its non-adaptive feature.

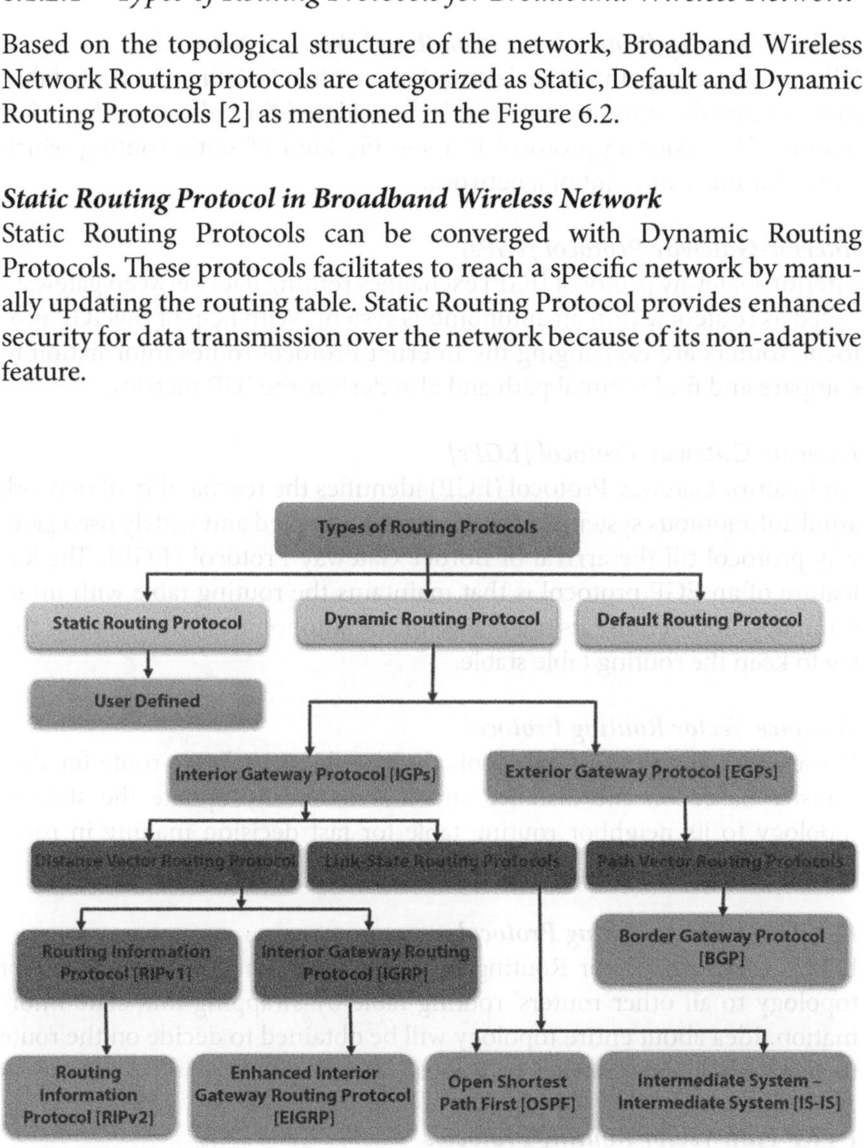

Figure 6.2 Wireless broadband networks routing-protocols.

DRP - Dynamic Routing-Protocol for Broadband Wireless Network

A Perfect routing decision is made dynamically by coping up with dynamic variation of topological state and update of inclusion and exclusion of network nodes. The routing mechanism is liable for calculating the superlative routes for diverse source-destination sets.

Default Routing-Protocols for Broadband Wireless Network

When no precise path to the destination terminal is defined in route table, then a general common route can be considered for delivering the information. This Routing protocol is a specific kind of static routing which works for one exit point of a network.

Interior-Gateway Protocol [IGPs]

Interior-Gateway protocol [IGP] exchanges routing data between gateways called as routers within an autonomous system. With help of this IGP protocol, routers are exchanging the Internet Protocol routes information to compare and find optimal path and also derives the IGP metrics.

Exterior-Gateway Protocol [EGPs]

An Exterior Gateway Protocol (EGP) identifies the reachability of network amid autonomous systems. EGP is an most accepted and widely used gateway protocol till the arrival of Border Gateway Protocol (BGP). The key feature of an EGP protocol is that maintains the routing table with minimum updates. Because its not reacts for the autonomous system issues and try to keep the routing table stable.

Distance Vector Routing Protocol

Distance Vector Routing protocols decides the superlative route for data transfer based on the distance and it dynamically update the state of topology to its neighbor routing table for fast decision making in route identification.

LSRP Link State Routing Protocol

LSRP Link State Vector Routing procedures dynamically update state of topology to all other routers' routing table by swapping link state information. Idea about entire topology will be obtained to decide on the route through which data transfer may occur [3].

PVRP Path Vector Routing Protocols

The main objective of the PVRP protocol is dynamically updates and the path information. In Path-Vector Routing Protocol, Path of the topology is actively updated in routing table along with next hop and destination address, depending on neighbor's router path for route identification.

RIP Routing Information Protocol

RIPv1, Routing Information Protocol is a classful routing, which doesn't include subnet mask in route advertisement. Broadcast Technique is used for intermittent updates [3].

RIPv2, this Routing Information Protocol is a classless routing, which carries subnet mask in route advertisement. Routing tables are updated and the information are exchanged only when there is an update in the topology.

Interior Gateway Routing Protocol [IGRP]

Interior Gateway Routing Protocol [IGRP] deals with the flow of path information within related routes in a host network, also known as autonomous systems (AS). Every switch's route tables are refreshed with the best possible approach or path, thanks to the convention. IGRP also avoids coordinating loops by resurrecting itself in response to group movements and identifying and recovering from loss or mistake.

Enhanced-Interior Gateway Routing Protocol [EIGRP]

EIGRP is a typical improved distance-vector path-finding for automatic routing discovery by active update of routing table and network topology.

Open Shortest Path First [OSPF]

OSPF identifies optimal route between source & destination by implementing link state routing protocol procedures.

Intermediate System – Intermediate System [IS-IS]

Intermediate System – Intermediate System [IS-IS] is an optimal information transfer within an autonomous network. It is an IGRP and much performs like link state routing.

6.2 Routing Mechanism in WiMAX Mesh Network

Wireless Mesh Network (WMNs) is an effective Mobile Ad-Hoc Network called MANET that establish, construct and reconcile itself dynamically with improved scalability and consistent services by balancing traffic in order to assure fixed or mobile connectivity. Extemporizing the standards of mesh network is from IEEE 802.11, IEEE 802.15 and IEEE 802.16 to

Wireless Fidelity (Wi-Fi), Bluetooth and WiMAX (Wireless Interoperability for Microwave Access) [4] respectively.

Wireless Communication Technology called WiMAX is purely meant for MAN-Metropolitan Area Network, which Consist of two parts namely WiMAX-Receiver and WiMAX-Tower as or Subscriber Station (SS) and Base Station (BS) respectively. IEEE 802.16 outlines flow of wireless traffic between subscribers and the base networks. WiMAX enables multimedia services such as video streaming and Voice Over Internet Protocol (VOIP) in mesh mode.

6.2.1 Routing Protocols - IEEE 802.16 Mesh Network

The IEEE 802.16 Mesh Network Routing Mechanisms is of two common types and they are Enhanced Distributed Routing Mechanism and Enhanced Centralized Routing Mechanism [5] as shown in Figure 6.3.

Enhanced Distributed Routing Mechanism

In Enhanced Distributed Routing Mechanism, every node collects network information from its next neighbor nodes. Based on collected information route to the destination is computed and through which traffic is forwarded. As it is a distributed Routing Mechanism, no crystal clear Base Stations defined in the network.

Enhanced Centralized Routing Mechanism

In Enhanced Centralized Routing Mechanism, Base Station performs the major task in assigning paths to all Subscriber Stations as in point to multipoint mode [PMP], and Multi-hop routing adapted Subscriber Stations relays the traffic from or to the Base stations.

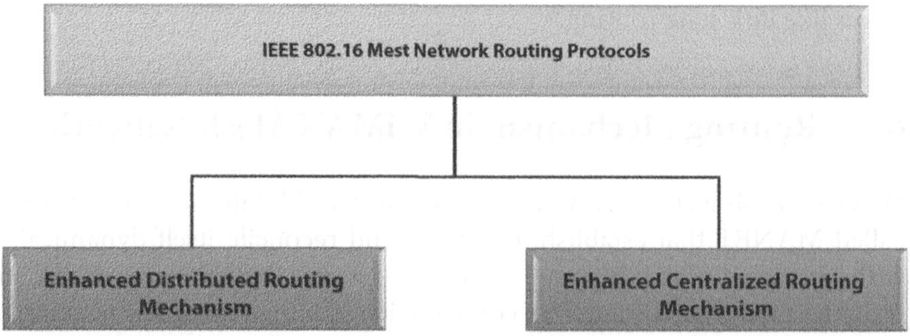

Figure 6.3 Routing protocols for IEEE 802.16 mesh network.

Robustness Issue in Wireless Mesh Network [WMN] Routing Mechanism

WMNs afford the most significant component for robustness in communication. Wireless networks are more error prone than wired network. Robustness is graceful degradation of traffic flow during error occurrences than that of total break in connectivity. This assures the improvisation in quality of service. Unlicensed frequency range usage by WMN provides graceful quality degradation which may work on frequency variety, by means of using multiple radio interfaces.

Quality of Service Issue in Wireless Mesh Network [WMN] Routing Mechanism

Quality of Service (QoS) in Multi-Hop WMN offers assured bandwidth for connection request or establishment. This Request or Establishment can be obstructed if a path cannot be found. The difficulties wing up when other routers interference affects the bandwidth for a router in the network. This encompasses QoS Issue.

Collision Avoidance Issue in Wireless Mesh Network [WMN] Routing Mechanism

As the number of subscriber station increases dynamically, usage of same bandwidth will also be increased much. This leads to major issue called collision and should be avoided by considering Collision metric value by Subscriber Station when entering the network.

6.2.2 Architecture of WiMAX Protocol

The WiMAX architecture is depicted in Figure 6.4. WiMAX Consist of two parts namely Base Station-BS and Subscriber Station-SS. The Subscriber Station may of Fixed or Mobile type. Central Node is the Base Station which coordinates all connection establishment along with routing to the Subscriber Station. The Router, backbone of WMN connects the subscriber through radio links.

There are two types of Subscriber Stations. They are defined as follows,

1. Indoor Subscriber Stations – Devices and connections used in indoor environments
2. Outdoor Subscriber Stations – Devices and connections used in outdoor environments

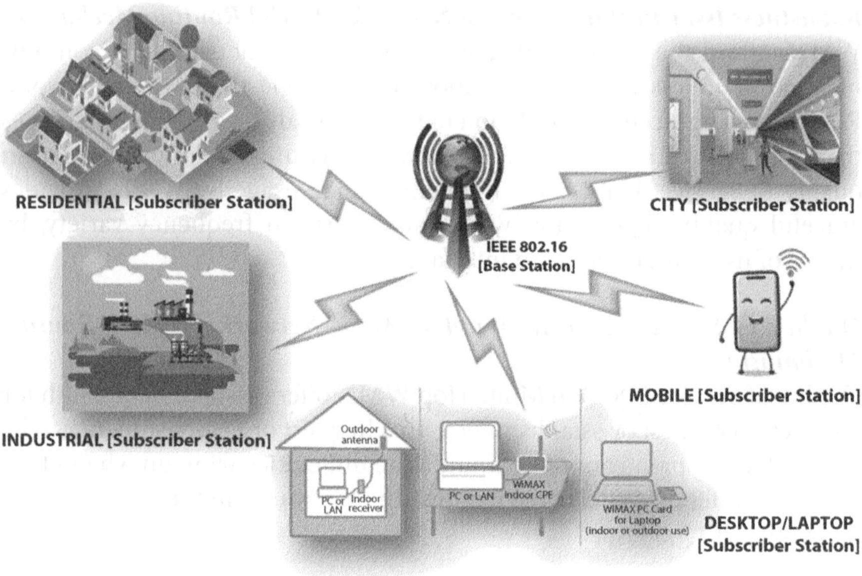

Figure 6.4 WiMAX architecture.

6.3 Routing Mechanism in Mobile Networks

6.3.1 4G Mobile Network

4G is generally called as fourth generation Mobile Networks. 4G network broadband cellular communication abilities are precise by ITU (International Telecommunication Union) a part of IMT Advanced (International Mobile Telecommunications Advanced). 4G Mobile communication networks works on four general routing Mechanisms and they are RIP, IGRP, EIGRP, Zebra in IP Routing Managers and OSPF [6].

Network Standards of 4G Mobile Network

The General network standards of 4G Mobile network [7] are

1. LTE – Long Term Evolution
2. LTE Advanced – Long-Term Evolution Advanced
3. Wi-MAX – Worldwide Interoperability for Microwave Access
4. UMB – Ultra-Mobile Broadband

LTE-Long-Term Evolution

It is as well called as Super-3G or 3.9 G which achieves very high speed with low packet latency. LTE Introduced Mechanisms like OFDMA-Orthogonal Frequency Division Multiple-Access and SC-FDMA-Single-Carrier Frequency Division Multiple-Access.

LTE-Advanced: Long Term Evolution Advanced [LTE-Advanced]

In LTE-Advanced, key necessities and key features numbers are stated.

1. Determined speed of downlink - 1Gbps
2. Determined speed of uplink - 500 Mbps
3. Latency <= 10ms
4. Efficiency of highest range for
   Downlink range -30bps/Hz
   Uplink range - 15bps/Hz
5. Supportable bandwidth - 100MHz

WiMAX

Mobile WiMAX belongs to 4G mobile communication, but all 4G mobile communication will not be mobile WiMAX. It covers a vast range of geographical space and also offers high speed internet access.

Ultra Mobile Broadband

UMB achieves higher data rate and increased speed with graceful degradation during loss of network and improvised Quality of Service. High data rate is obtained by the usage of technologies like TDM [Time Division Multiplexing], CDMA [Collision Detection Multiple-Access] and OFDMA [Orthogonal Frequency Division Multiple-Access].

6.3.2 5G Mobile Network

TSP-Telecom Service Provider practices radio frequency range for data communication. As Mobile usage population increases day by day, data consumption of the same frequency range also gets increased. This increases less bandwidth allocation to the subscribers and which leads to slow connection or even connection loss. To answer such issue 5G Mobile Network is used to provide lower cost high data rate [8]. 5G is beset to resolve inconsistent demands during the year 2020. 5G is also referred as was labelled as Ultra Technology as 'Ultra-Fast', 'Ultra-Maximum Capacity' and 'Ultra-Reliable' with minimal latency. Figure 6.5, represents the Architecture Diagram for 4G-5G Mobile Network.

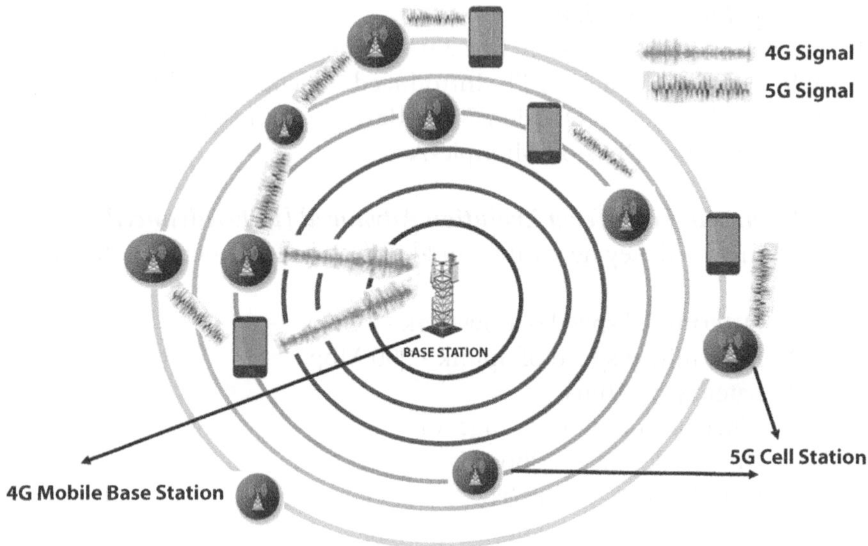

Figure 6.5 4G – 5G mobile network architecture.

6.3.3 Convergence Communication Issue

Why from 4G to 5G?

Figure 6.6 discusses the reasons for moving from 4G to 5G Mobile network based on Latency, Traffic flow, Greatest Data Speed, Existing Range and Connection Mass.

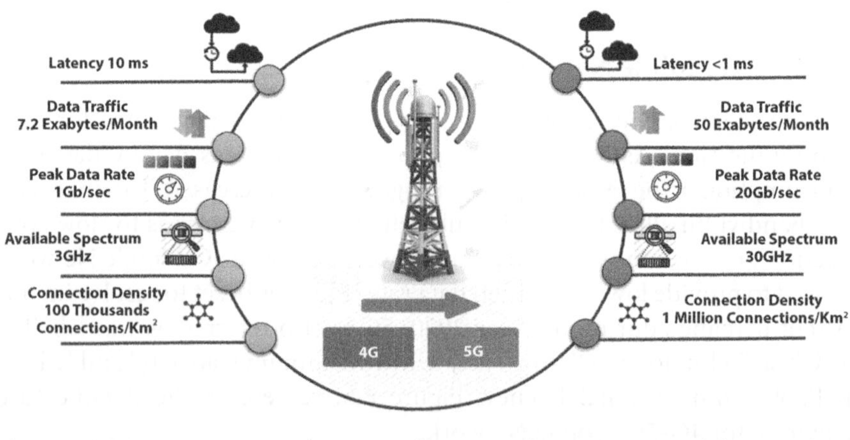

Figure 6.6 Diagram represents the reasons for moving from 4G to 5G mobile network.

Issues in convergence communication are the challenges faced during the process of converging two different kinds of communication technologies like 4G which gets converged with 5G network. The fourth generation mobile network deploys to advanced telecommunication technologies like 5G technologies and its services [9].

Similarly, the terminals of the network should also be in cope up with the new policies and the challenges of the new technologies. As the 5G mobile network has the improved standards like higher peak data rate, more data traffic, wide range of available spectrum, lesser latency and higher connection density but in 4G the mentioned features are much lesser in performance than that of 5G and there is a need for converging technologies to enhance the 4G network to combine with 5G network to improvise the performance of the mobile network.

6.3.4 Multi-Hop Dynamic Routing Issue

The performance of the 4G mobile network suppressed by shadowing and multipath issues. The capacity and the coverage of the mobile networks are enhanced by RS- relation stations in multi-hop relay networks.

Route identification and relay stations configurations are the major issues which suppresses the QoS-Quality of Service of mobile network. In order to optimize QoS-Quality of Service of the mobile network, route identification can be done using LARM-Load Aware Routing Metrics and BP-Burst Profile [10] to enhance relay stations Identification.

LARM-Load Aware Routing Metrics

LARM considers the load of local traffic and the interference to assure load balancing in wireless networks. The load of local traffic can be identified in two different ways and they are as follows,

1. Identifying the traffic of live node.
2. Identifying the traffic of peer node

The live node traffic identification represents the active part of live node in the network. The Peer node traffic identification represents the busyness of the peer nodes in the network.

BP-Burst Profile

Every SS-Subscriber Station assigns a burst profile with the base station. Burst profiles are chosen dependent on QoS-Quality of Service needs and channel conditions. The burst profile is an essential device in the 802.16 MAC Layer Standard. The burst profile distribution, which changes powerfully and potentially quick, is about actual transmission. Here the boundaries of the burst profiles of WiMAX are summed up. The burst profiles are utilized for the Link adaptation methodology. There are two types of Burst Profile and they are defined as follows,

1. Downlink Burst Profile – Subscriber Station creates Request for New DL Burst Profile
2. Uplink Burst Profile – Base Station creates Request for New DL Burst Profile

6.4 Service Specific Routing Protocols

The aim of service-specific routing protocols is to make it easier to create an innovative route among sender and a target abided with minimum one path among them. A protocol developed for routing ought to do two tasks to complete the choices: (1) choose the optimal course based over its metrics of routing, and (2) promulgate knowledge of network behavior used for route choice. The classification of Service Specific Routing Protocol is represented in Figure 6.7.

6.4.1 Protocols Based on Topology and Position

Protocols for topology-based routing choose pathways stands on geometrical data like node connection condition and length of routes. This set of protocols on routing, according to the literature, involves the establishment of a route until a message can be transmitted from a sender to its destination network. Ad hoc distance vector (AODV) [11], optimised connection state routing (OLSR) [12], and dynamic source routing (DSR)

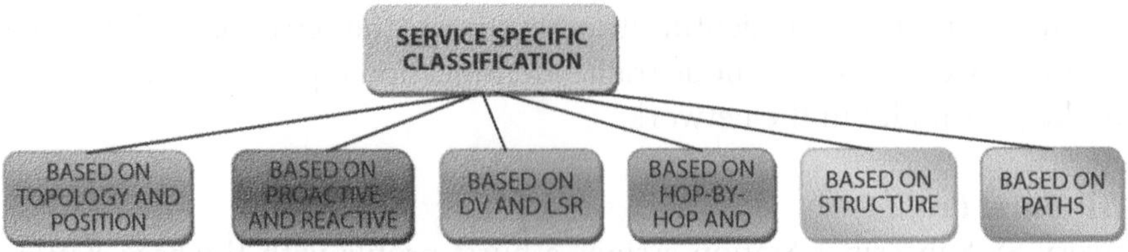

Figure 6.7 Service specific routing protocol classification.

[13] are examples of routing protocols which is topology-based. Routing protocols based on position [14], unlike routing protocols which is based on topology, do not entail the invocation among a route between a initiator address and its recipient node until the sender transfers the message. Alternatively, it allows information-by-information forwarding choices, and successive packets in multiple choices can be routed down by multiple separate routes.

The spatial locations of the heeding node, assisted neighbours, and the recipient system are used to make the packet-forwarding decision. Geographic routing protocols are also known as routing protocols based on position. Nodes have been focused by their structural or spatial locations rather than their network addresses in position-based protocols. A location based service device, which will be a key aspect of the routing protocol, provides information about position of subsequent nodes.

An algorithm which denotes the common forwarding mechanism can be used to decide the nearest distance among every messages based on the location information, with the nearest hop being chosen so that the message is nearest to its target node. Regretfully, despite the fact that a path to the destination exists, this basic algorithm can impart the data packets to become interrupted in a minimum local threshold and hence destination is not permanently reached. Because of the local minimum dilemma, a rollback recovery solution for forwarding the jamming information to be delivered out is needed. To implement this recovery, an approach is found that divides the cluster into several disseminates, each with its own set of connections that do not cross. By moving through these faces into the destination, packets will get out of a local minimum [15]. Location-aided routing (LAR) [16] and greedy perimeter stateless routing (GPSR) [15] are two examples of position-based routing protocols. GPSR blends aggressive forwarding and fallback contact forwarding.

6.4.2 Protocols Based on Proactive and Reactive Nature

Table-driven routing protocols are another name for proactive routing protocols. Even if the routes maintained are not expected or in service, this class of protocols of routing maintains a route in the network which is distributed to other nodes. Routing control messages are sent on a regular basis to check the connection category and tables which holds the routing information by the nodes, ensuring that all routes are kept up to date. As a result, when a new path is required, there is no route acquisition latency.

However, maintaining all available routes' routing information at all times will result in a higher network load .This is inefficient, particularly because most of the directions kept have never been used. The frequency of periodic control messages is important in constructive routing protocols for reliably capturing network dynamics. When the upgrade frequency isn't sufficient level, a stale or redundant route might be used, ending in dropped packets. OLSR [12], topology transmitted based on reverse-path forwarding (TBRPF) [17], and DSDV [18] are examples of proactive routing protocols.

Routing protocols which is based on ad-hoc nature can be called with another name for reactive routing protocols. Periodic control messages are not sent by reactive protocols to change connection states with all possible routes. Rather, the computation is done whenever it is needed, and it barely keeps track of routes that are actively being used. While only a tiny proportion of all possible paths is in operation at any given time, it eliminates power overhead. Since a elevated transparency shows a major performance effect in bandwidth which has low computation wireless connections, the diminution of control routing transparency stays at the primary inputs following the implementation of protocols which is based on ad-hoc nature. Nevertheless, since a route discovery mechanism is worn to group a innovative pathway ahead of a data packet be able to sent, worn to group a pathway induces on-demand acquisition route latency. When the direction is being set up, the route acquisition complexity may cause a significant delay for the preliminary messages waiting to be sent. The routing algorithm based on temporally ordered (TORA) [19], DSR and AODV, are examples of routing protocols which are reactive in nature. Figure 6.8 depicts the comparison analysis of routing control overhead with respect to reactive and proactive types.

6.4.3 Protocols Based on Distance Vector and Link State Routing

Routing protocols based on Distance vector enables each device to transmit its distance among other nodes within the network to all of its immediate neighbours. A node can choose the shortest path to each node based on distance information obtained in the network abided among neighbours. It is not appropriate for neighbour node broadcasts to contain only distance information. In reality, in addition to distance, the broadcast can include metrics on routing such as CETX, CETT (path length or hop count) and CRTT. Such routing metrics will also be used for shortest path coverage by weighing the values. Constituent systems of this group of protocols

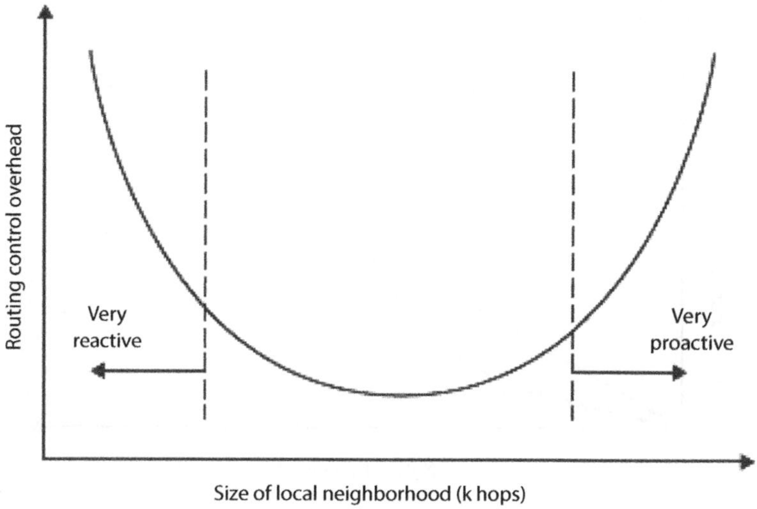

Figure 6.8 Comparison of routing control overhead in reactive and proactive [21].

on routing are most often referred to as route vector routing protocols because the metric or parameters of routing is not necessarily be length of path. AODV and DSDV are two distance (path) vector routing protocols.

According to routing protocols based on link state, all nodes compulsorily transmit about their status on links not only among instantaneous neighbours, but also among all nodes inside the cluster. As a result, based on the most current connection in sequence among all other nodes, each node can create a complete image of the network. A nodule node will desire the shortest path among other nodes in the association if it has all of the network's knowledge.

OLSR is an example of a routing protocol which is based on connection state. However, the OLSR protocol televise only to immediate neighbours with distance information, routing protocols based on distance vector can hold upon a subordinate control overhead on routing than routing protocols based on link state. The routing overheads of distance vector and connection state routing protocols are compared in Figure 6.9.

6.4.4 Protocols Based on Hop-by-Hop Routing and Source Routing

In routing protocols based on Hop-by-hop, all nodes must hold its own table information on routing data The nearby hop that holds routes among

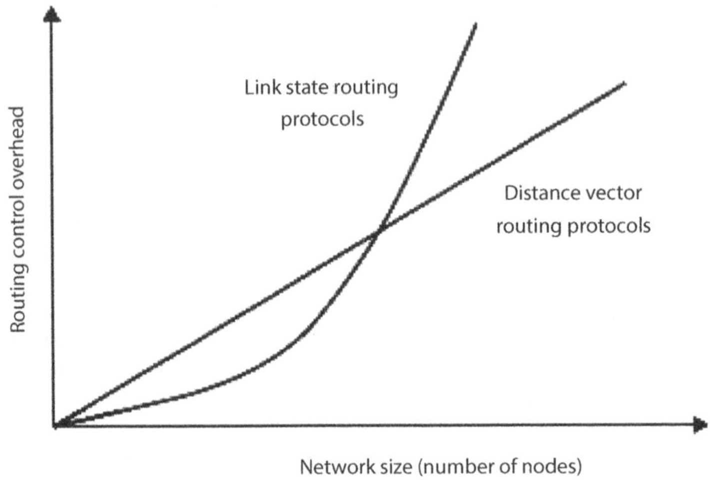

Figure 6.9 Comparison of routing control overhead in DV and LSR [22].

other nodes in the group is shown to inculcate the distances. The destination address is all that is needed for a package to accomplish its intentional destination. The data packet is merely forwarded along its way through intermediate nodes depending on the destination. Routing protocols based on hop-by-hop are common in Adhoc networks and networks based on wireless mesh because of their quick forwarding and low message overhead. AODV and OLSR are examples of hop-by-hop routing protocols.

Routing tables are not needed at all intermediate nodes for routing protocols based on source paths. Instead, the source node is the only one that distinguishes the whole path based on hop-by-hop to the target node. At the root system, the nearby routes have been estimated and saved in a path cache. The initiator node removes the route of source from its routing updates and places the intact pathway in the subtitle header before transmitting packets. nodes which are intermediate cannot use table information of routing; instead, they must forward packets with respect to route sequences inside the header information of packet. This result in huge incoming packets, which makes a channel of wireless in nature based on the error rate makes more fragile. DSR is one example of a root routing protocol.

6.4.5 Protocols Based on Flat and Hierarchical Infrastructure

Many of the routing protocols we've encountered so far in this segment presume architecture with respect to flat network, in which all peers are considered as nodes with one or more hops between them. Each node in this architecture with respect to flat network is required on to aware of network topology modification that affect its routing table. However, based on the routing protocol characteristic based on constructive or imprudent, as the network topology changes regularly, a significant amount of routing control traffic is created. As a result, these routing protocols based on flat architecture are adaptable to modifications on the number of systems, connection efficiency, and node mobility in the network.

Layers with respect to hierarchy can be implied on a multihop broadband wireless network, in contrast to a flat network architecture, where nodes based on subset in nature have been chosen to cluster a virtual backbone wireless network. Certain other modules that aren't connected to the immersive backbone can communicate through it. As the network topology changes, a non-backbone link often needs to notify its closest backbone link, while a link of backbone merely needs to notify all nodes with respect to backbone structure. As a result, when opposed to a flat routing protocol, routing control overhead can be significantly reduced [20].

Moreover holding of all the systems that have a narrow range of transmission, the possibility to choose for a heading information of cluster to be linked to other clusters only holding the head nodes with respect to non-cluster of the hierarchical infrastructure. The gateway node denotes the head node with respect to non–cluster that connects multiple neighbouring clusters in this scenario. Nodes which act as gateway ensures that every group holding the network is linked.

6.4.6 Protocols Based on Single-Path and Multipath

For each pair of foundation and target nodes, only one end-to-end route is preserved among entire protocols of routing mentioned so far. Such single-path routing protocols become distinct from routing protocols that maintain many routes based on end-to-end paths. The protocols which works based on multipath routing can help a multihop wireless broadband network improve its end-to-end efficiency. When a route is interrupted, for instance, in multipath protocols which works based on multipath routing, alternative paths can induce automatically to assist for transmissions with abiding end-to-end delays, preventing a examined interruption. One of the examples for multipath routing is AOMDV (ad hoc on-demand multipath distance vector routing) protocol.

6.5 Novel Approaches and Algorithms on Broadband Routing

6.5.1 Approaches for Efficient Performance in Broadband Routing

Cost setup:
The different destination nodes start this cost setup process. It allows every node to set it's own cost in relation to every targeted host. The minimal quantity of hop distance required to reach a target node is represented based on the cost of a node against that destination network. The cost configuration will aid the exploration agents in finding routes to the receiver nodes.

Route discovery agents:
Every sensor in the network sends out agents to search for paths to the destination host. Those participants can use the expense to travel the network in a randomised manner. They would normally head in the pursuit of lower expense, but they would sometimes choose an access vital direction. Random nature is important because it allows you to discover additional routes and avoid being "stranded" in global minimum. Consider what would happen if exploration officers just went in the direction of lower costs. If a node with a cost of 3 had only one neighbour with a cost of 2 who expired, no routes will ever be identified from such a route. As a result, the aim of this approach is to find as many pathways as practicable.

Maintenance agents:
Maintenance agents are sent back from the destination nodes to offer feedback on the found routes in terms of latency and distribution performance %.

Feedback Process:
The input function would choose a selection of available data transmission routes on a probabilistic basis. The outcome can be made based on the success rate of every route's execution. The link with the lowest packet distribution ratio will calculate this rate, and maintenance agents then

communicate it to the originated nodes. This evaluation process is critical because it provides source nodes with an up-to-date picture of each route's output at any given time. As a result, they'll be capable of directs the data using most efficient route.

6.5.2 Algorithms for Broadband Routing

The algorithms can be classified such as power-aware routing, flat routing, location-based routing and hierarchical routing.

In power aware routing, the inconsistency of the connections, as well as the limits of all tools, make routing extremely difficult. Also in the existence of access links, channel conditions change due to rayleigh propagation effects [23] caused by sensors or objects moving about in the atmosphere, which alter the patterns of radio frequency vibrations. Furthermore, since sensing nodes are normally battery-powered, continuous monitoring would not be feasible: the gradual depletion in usable resources must be taken into account. The key factors that form the network graph are the strength of the connections and the transmission power in the nodes; connection quality can be determined unswervingly by some of frequencies, while vitality consumption is connected around the bridge level of voltage that can be calculated and inserted through into microcontroller. Moreover the figures can be combined to provide an evaluation computation for determining the most effective path. Furthermore, to lessen the delay, the existence upon time limit forces the association of network to prefer common routes with a limited values of single hop distances (a.k.a. the solution based on long-hop).

As a result, a cross layer cost function is calculated, that involves pure device information (residual energy), core network data (quality in connection), holding with navigation layer variable (hop values). The approach determined by long-hop allows the sharing information of consistent time of the crucial region by utilization of energy of nodes; intuitively, a scheme based on route selection concepts the performance of energy management of nodes which helps in lifetime extension of entity nodes and for the accomplishment of nodes energy balance in the set of connected and grouped systems.

Flat Routing

Every node in flat routing has the same duty, i.e. each node participates in the information routing. Some of the examples of flat-routing are AODV, DSR, Directed Diffusion, rumor routing.

Both unicast and multicast routing are possible with the Ad-Hoc On Demand Distance Vector (AODV). It's a on demand method [24], which means it only creates routing among nodes when source nodes need them. A path request packet (RREQ) is transmitted through the network whenever a node have to chose a pathway to transmit data to the destination for which it is not yet chosen a route. Pointers with respect to nodes set up for the destination, the source node gets back as a response for the route request (RREP). The node will start sending data to the destination until it receives the RREP. AODV, on the other hand, mandates that contact be postponed before a path to the destination node is revealed.

Dynamic source routing

A sender which sends the packets will decide the sequence of the intermediate nodes where the data packet is routed sequentially would attains target node in dynamic source routing (DSR) [25]. The route's succeeding node ID will be included on the header of every packet. Suppose if the succeeding node is not a destination node, then the received message will forward toward the next hop on the determined path and it continues the same routine till the data packet reaches the destination node. To enable the dynamic routing mechanism an discovery protocol helps the node to find an alternate path and finds an route with another node in the network on the move.

Directed diffusion

For broadband wireless networks, directed diffusion as a data-centric, data distribution model. Nodes that seek information are known as sinks, whereas nodes that create information are known as sources. Interests are entries that show a desire for a specific type of data [26]. Sinks will create interests, which will spread across the network in search of nodes that have event histories that fit. Events begin to flow in several directions through the originators of interests. An interval attribute field in an interest specifies how often data relevant to that interest should be submitted. During this time of contact, the sensor networks fortify any one or some paths across the course of their usage. Because it may contact a target node via a variety of paths, this suggested system is immune to node failures, but it increases traffic by overloading the network with requests. The generic architecture [27] of the diffusion algorithm as shown in Figure 6.10.

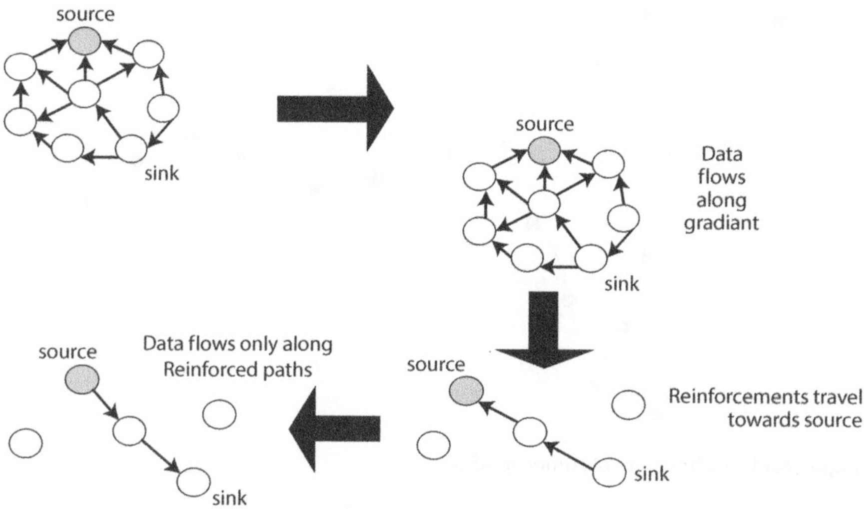

Figure 6.10 Directed diffusion approach.

Rumor routing

Rumor routing algorithm in WSN is a proven successor of directed diffusion. Certain cases observed that the exchanged total data is insignificant. Though sink nodes send requests will be a 2D (multi routes) fashion could meet originating nodes with the guided diffusion algorithm, all sink nodes and source nodes scatter messages in a single-dimensional (random walk) manner in the gossip routing algorithm (one dimensional route). A direction between the two has been formed when the curve originating from a source approaches the curve emanating from a corresponding drain [28] [29]. By avoiding overwhelming the network with requests, event alerts, or differential generated messages, this algorithm decreases the amount of interaction. Near-optimal routes and high bandwidth requirements have a tradeoff. Figure 6.11, depicts the architecture of rumor routing.

Location based routing

Location-Based Architecture is therefore reliant on the knowledge contained in the nodes that it encounters. The position of nodes is determined with the help of low-power GPS receivers, either by the Triangulation Approach or by placing them in a known location. The styles of Location-Based Architecture are defined in the following classification. Geographic Adaptive Fidelity (GAF) [30, 31] is a merged category of energy whose orientation is determined by different levelled models. It saves resources by turning off redundant nodes and then locating equivalent nodes for

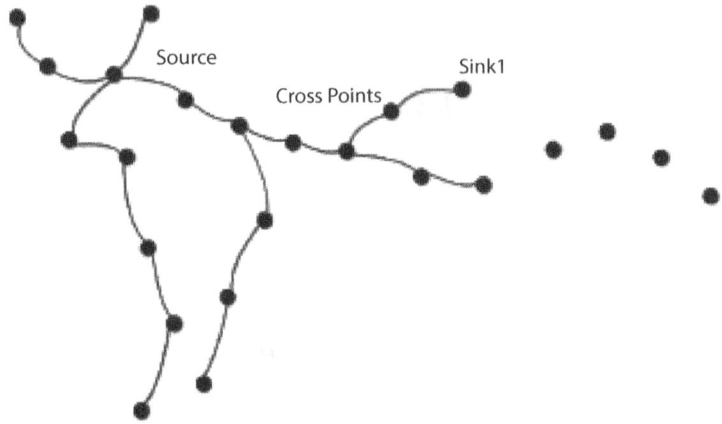

Figure 6.11 Architecture of rumor routing.

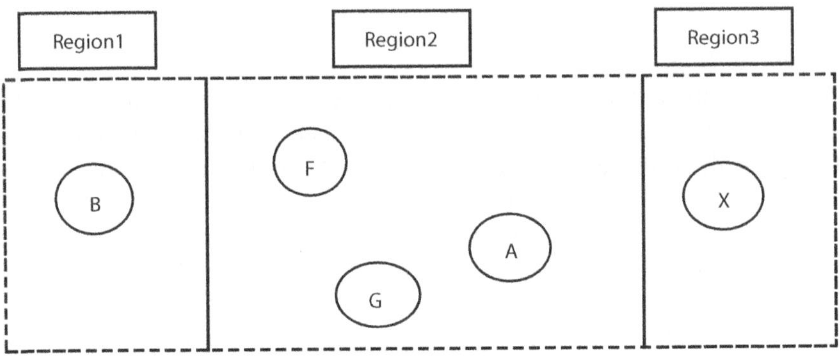

Figure 6.12 Architecture of GAF routing.

routing. It maintains its routing and vitality utilization at this stage. Figure 6.12 depicts on the architecture of GAF routing.

Geographic Random Forwarding (GeRaf) is based on a self centered model, in which it selects a node from the vicinity of the destination node and creates a path to its destination. First, an extremely well-organized area is chosen based on the position knowledge of nearby nodes. At this stage, a Request-To-Send (RTS) message is sent to all nodes within the specified area. If the priority area sends a Clear-to-Send (CTS) message, the initiator node either transmits the packet or continues investigating the priority region before it receives a CTS message [32, 33].

Geographical And Energy Aware Routing (GEAR) is very specific for it is aware about the energy routing system that chooses a neighbour to direct requests to the destination region. Usually all routing algorithms will directs the packet to the specific node, but the GEAR routing protocol will transmits the packet to the destination region. It requires a Geographical Information System or region locate system to know location information because this protocol mainly based on the location. By regulating energy, this protocol will reduces delay and extends lifespan [34].

Hierarchical Routing

In hierarchical routing, not every node has the same feature. Information which will gathered through sensing nodes will be further transmitting to the base station is the responsibility of certain unique nodes or cluster-heads. As a result of the more difficult connectivity limitations, cluster-heads use more resources than sensed data nodes. The global routing techniques has been optimized to achieve effective VLSI physical layout design [35].

Low Energy Adaptive Clustering Hierarchy (LEACH) [36] technique involves with adjusting an network lifetime in an arbitrary way and thus distributing the energy stack across sensor base stations. It also allows sensor nodes to choose their cluster head on their own. It necessitates the least amount of energy in this case. This results in a shortage of cluster formation at the appropriate time load. It also eliminates the need for cluster head swapping which allows for more efficient use of the system's capacity cap due to the elimination of the overhead associated with swapping.

The Threshold Sensitive Energy Efficient Sensor Network (TEEN) [37] is based on a two-stage configuration of Cluster Head (CH) nodes and Separate Nodes. CH combines the data from all of the nodes connected to it and sends it to the base station. When one cluster has an immediate uplink path to at least one CH from the root, a second-level cluster head is created. The correspondence of a single node to its CH is determined by a hard limit (ie: if an explicit value is achieved, then it should be informed to CH). The delicate threshold depicts the set delta value that must be increased. If there is a need for occurrences provided as occasional, TEEN is unprepared to achieve desired results.

In Virtual grid architecture routing, by grouping nodes in standard non-overlapping clusters, the network is separated into rectilinear grids [38, 39]. A node is optimally chosen to serve as the cluster-head within each cluster. Data processing occurs on two occasions. To begin, each cluster-head, also known as a Localized Aggregator, is responsible for performing a localized data processing. Until sending the data to the base station, cluster heads branches will performs the aggregation at global level. The grid that has been built should be shifted to ensure proper energy balance.

6.6 Conclusion

In this book chapter, various types of routing protocols and its mechanisms for WiMAX based IEEE 802.16 Mesh networks and mobile networks have been discussed. WiMAX routing is a dynamic space of research wherein many investigations and proposals are in progress for increased data rate, minimized latency, maximum throughput, improved Quality of Service and robustness through wireless broadband networks. The issues like convergence communication, Multi-Hop dynamic routing and other specific issues are also summarized based on a survey of recent literature articles. Also this chapter emphasizes routing protocols used for some specific application. Analysis and discussion of broadband innovative approaches and future implications of routing protocols are discussed.

References

1. Rückert, J., Blendin, J. & Hausheer, D., "Software-Defined Multicast for Over-the-Top and Overlay-based Live Streaming in ISP Networks," J Netw Syst Manage, Vol. **23, pp.** 280–308, 2015. https://doi.org/10.1007/s10922-014-9322-8.
2. H. Karna, V. Baggan, A. K. Sahoo and P. K. Sarangi, "Performance Analysis of Interior Gateway Protocols (IGPs) using GNS-3," *2019 8th International Conference System Modeling and Advancement in Research Trends (SMART)*, pp. 204-209, 2019.doi: 10.1109/SMART46866.2019.9117308.
3. Sun, C., Yin, B., Dou, Z. *et al.*, "A Routing Protocol Combining Link State and Distance Vector for GEO-GEO Satellite Backbone Network," Mobile Netw Appl Vol. 24, pp. 1937–1946,2019. https://doi.org/10.1007/s11036-019-01339-y.
4. Tássio Carvalho, José Jailton Júnior, Warley Valente, Carlos Natalino, Renato Francês and Kelvin Lopes Dias (June 5th 2013). A Mobile WiMAX Mesh Network with Routing Techniques and Quality of Service Mechanisms, Selected Topics in WiMAX, Gianni Pasolini, IntechOpen, DOI: 10.5772/55863. Available from: https://www.intechopen.com/books/selected-topics-in-wimax/a-mobile-wimax-mesh-network-with-routing-techniques-and-quality-of-service-mechanisms.

5. Qasem, Yaaqob & Al-Humairi, Ali & Ng, Chee Kyun & Noordin, Nor & Rasid, M, "Review of network routing in IEEE 802.16 WiMAX mesh networks", Australian Journal of Basic and Applied Sciences. 3, No.4, pp: 3980-3996, 2009.
6. IKram Ud Din, Saeed Mahfooz and Muhammad Adnan, "Analysis of the Routing Protocols in Real Time Transmission: A Comparative Study", Vol. 10, No. 5,2010
7. Shoewu, Oluwagbemiga., "Evolution of the Fourth Generation Mobile Networks: Trends and Tendencies", Vol. 8, No. 2, pp. 392-398, 2018.
8. Xiang, W., Zheng, K., & Shen, X. (Eds.). (2017). *5G Mobile Communications.* doi:10.1007/978-3-319-34208-5
9. Toral-Cruz, H., Mihovska, A.D., Gaj, P. *et al.*, "Advances and Challenges in Convergent Communication Networks", *Wireless Pers Commun*, Vol. 96, pp. 4919–4927, 2017, https://doi.org/10.1007/s11277-017-4964-y.
10. Le, Anh-Ngoc *et al.* "LARM: A Load-Aware Routing Metric for multi-radio Wireless Mesh Networks." *2008 International Conference on Advanced Technologies for Communications*, pp. 166-169, 2008.
11. C. E. Perkins and E. M. Royer, "Ad-hoc on demand distance vector routing," *Proceedings of the IEEE Workshop on Mobile Computing Systems and Applications*, pp. 90–100, Febr. 1999.
12. P. Jacquet, P. Muhlethaler,T. Clausen, A. Laouiti, A. Qayyum, andL.Viennot, "Optimized link state routing protocol for ad hoc networks," *Proceedings of the IEEE Multi Topic Conference*, pp. 62–68, 2001.
13. D. B. Johson and D. A. Maltz, "Dynamic source routing in ad hoc wireless networks," in *Mobile Computing*, Kluwer Academic, Norwell, MA, Chap. 5, 1996.
14. M. Mauve, J.Widmer, and H. Hartnstein, "A survey on position-based routing in mobile ad hoc networks," *IEEE Network*, vol. 15, no. 6, pp. 30–39, Nov. 2001.
15. P. Bose, P. Morin, I. Stojmenovic, and J. Urrutia, "Routing with guaranteed delivery in ad hoc wireless networks," *ACM Wireless Networks*, vol. 7, no. 6, Nov. 2001.
16. Y.-B. Ko and N. H. Vaidya, "Location-aided routing (LAR) in mobile ad hoc networks," *ACM/Baltzer Wireless Networks*, vol. 6, no. 4, pp. 307–321, 2000.
17. B. Bellur, R. Ogier, and F. Templin, "Topology broadcast based on reverse-path forwarding (TBRPF)," IETF Internet Draft Document draft-ietf-manet-tbrpf-01.txt, Mar. 2001.
18. C. E. Perkins and P. Bhagwat, "Highly dynamic destination sequenced distance vector routing (DSDV) for mobile computers," *Proceedings of ACM SIGCOMM*, pp. 234–244, Oct. 1994.
19. V. D. Park andM. S. Corson, "A highly adaptive distributed routing algorithm for mobile wireless networks," *Proceedings of IEEE INFOCOM*, pp. 1405–1413, 1997.

20. J. Sucec and I.Marsic, "Hierarchical routing overhead in mobile ad hoc networks," *IEEE Trans. Mobile Comput.*, vol. 3, no. 1, Jan. 2004.
21. P. Samar, M. R. Pearlman, and Z. J. Haas, "Hybrid routing: the pursuit of an adaptable and scalable routing framework for ad hoc networks," in *The Handbook of Wireless Ad Hoc Networks*, CRC Press, Boca Raton, FL, pp. 245–262, 2003.
22. M. Audeh, "Metropolitan-scale Wi-Fi mesh networks," *IEEE Comput.*, vol. 37, no. 12, pp. 119–121, Dec. 2004.
23. J.N.Al-Karaki and A.E.Kamal. "Routing Techniques in Wireless Sensor Networks: A Survey," IEEE Wireless Communications, Dec 2004.
24. R. Brockett, "Minimum Attention Control," Proceedings of the 36th Conference on Decision and Control, 1997, pp. 2628-2632.
25. V.Barghavan B. Das. "Routing in ad-hoc networks using minimum connected dominating sets," In IEEE International Conference on Communications(ICC), 1997.
26. E.Bonabeau,Marco Dorigo and G.Theraulaz."Swarm Intelligence From Natural to Artificial Systems" New York Oxford, Oxford University press, pp. 25-107, 1999
27. D.Braginsky and D.Estrin. "Rumor routing algorthim for sensor networks," In WSNA "02: Proceedings of the1st ACM international workshop on Wireless sensor networks and applications, pages 22–31. ACM Press, 2002.
28. H.Balakrishnan W.R Heinzelman A. Chandrakasan. "Energy-efficient communication protocols for wireless microsensor networks," In Proceedings of the Hawaian International Conference on Systems Science, 2000.
29. P. Bose and P. Morin, "Competitive online routing in triangulations," In Proc. 10th International Symposium on Algorithms and Computation (ISAAC, 99), pages 113-122. Springer-Verlag, LNCS 1741, 1999
30. H.Balakrishnan R.Morris B.Chen, K. Jamieson. "Span: An energyefficient-coordination algorithm for topology maintenance in ad hoc wireless networks," In ACM/IEEE International Conference on Mobile Computing and Networking(MOBICOM 2001), 2001. 77
31. D.S.J.D.Couto, D.Aguayo, J.Bicket, and R. Morris. "A High Throughput Path Metric for MultiHop Wireless Routing," In Proceedings of the 9th Annual International Conference on Mobile Computing and Networking (MobiCom'03), San Diego, CA, Sept. 2003.
32. D. Estrin Y. Xu, J. Heidemann. "Adaptive energy-conserving routing for multihop adhoc networks," In USC/ISI Research Report 527, 2000. ISCO 2008 on 1-2, February 2008 at KCE, Coimbatore - 32
33. N. Elia and S. Mittler., "Stabilization of Linear Systems with Limited Information," IEEE Transactions on Automatic Control, 2001, pp. 1384-1400.
34. J. Grover, Shikha and M. Sharma, "Location based protocols in Wireless Sensor Network - A review," Fifth International Conference on Computing, Communications and Networking Technologies (ICCCNT), 2014, pp. 1-5.

35. Dr.T.Rajendran, O.Vignesh, "Optimized Global Routing Techniques for Efficient VLSI Physical Layout Design", International Journal of Advanced Science and Technology, Vol. 29, No. 6, (2020), pp. 2570 – 2778 ISSN: 2005-4238, May 2020.
36. D. Estrin Y. Xu, J. Heidemann. "Geography informed energy conservation for ad hoc routing," In Proceedings of the Seventh AnnualACM/IEEE International Conference on Mobile Computing and Networking (ACM Mobicom), 2001.
37. E. N. Gilbert. "Random plane networks," Journal of SIAM, 9:533{543, Oct. 1961.
38. M. Gunes, U. Sorges, I. Bouazizi, "ARA - The Ant-ColonyBased Routing Algorithm for MANETs," Proceedings of the ICPP Workshop on Ad Hoc Networks (IWAHN 2002), IEEE.
39. W. Heinzelman, A. Chandrakasan, and H. Balakrishnan, "Energy-efficient communication protocol for wireless microsensor networks," Proceedings 33rd Hawaii Int. Conf. Syst. Sci. (HICSS "00), January 2000.

7

Interference Problem in 5G with Radio Access Network

S. T. Deepa* **and Vinitha Kumari**

School of Computational Science, Shri Shankarlal Sundarbai Shasun Jain College for Women, Chennai, India

Abstract

As network operators navigate a multi-pronged transformation that encompasses 5G, fiber proliferation, increasingly cloud-based and virtualized networks, one of the challenges that they face is that of radio frequency interference. RF interference is not a new issue in wireless networks, and in fact, some of the same sources of RF interference that affect 4G networks will also impact 5G networks. Here are some of the most important things to know about RF interference in 5G networks. 5G can be affected by the same RF interference as 4G because it may operate in the same frequency bands as 4G. These bands, whether they be the low, the mid or the high-frequency bands, they can be affected by sources of interference—and these sources of interferences are increasing everywhere, and have really become a huge headache for network operations teams. In the low and mid-bands where 3G, 4G and 5G may all be currently operating, he added, EXFO's customers are seeing tens of thousands of sites impacted by RF interference and specifically, passive inter-modulation or PIM. 5G Non-Standalone networks depend on 4G, so the same RF interference that impacts 4G will affect 5G. While it may be tempting to think that 4G issues wouldn't affect new 5G networks, Lipari points out that when it comes to 5G Non-Standalone—which is how most operators began, and are still, deploying 5G—the 4G network is an integral part of 5G, with control traffic traveling over the 4G uplink. In 5G NSA, Lipari added, operators need to pay attention to uplink issues in particular, as reflected in Received Signal Strength Indicators (RSSI)—because 4G uplink issues will impact the 5G network. RF interference impacts the utility of an operator's spectrum investment, and customer experience. He also noted that as channel bandwidths increase, from 5-20 megahertz in 4G to 100-400 megahertz in 5G, the number of resource blocks increases as well. If a resource block is has interference, Lipari explains, a user most likely will not be able to able to connect or will have bad signal quality. Operators spend billions

**Corresponding author*: stdeepa@shasuncollege.edu.in

A. Suresh, J. Ramkumar, M. Baskar and Ali Kashif Bashir (eds.) Resource Management in Advanced Wireless Networks, (141–160) © 2023 Scrivener Publishing LLC

of dollars on spectrum, and if it is impaired by interference, that reduces the utility of their investment. RF interference and PIM can create issues such as reducing the overall speed and capacity of a site, or it can rise to the level of causing unstable connections and dropped calls. A RAN performance engineer will analyze the relevant KPIs and look at historical statistics to determine the location of the problem, and then a ticket is passed to a crew who will visit the site—usually meaning that a cell tech will go and take measurements that will hopefully illuminate the issue.

***Keywords*:** Network operator, 5G non-standalone, RF interference, intermodulation, PIM, RSSI, RAN, bandwidth

7.1 Introduction

The growing explosion and demand for broadcasting emerged as the progress of the standard Radio Access Networks (RANs) seemed to be closer to the connected and unexpected distribution as it could be predicted on a number of five (5G) domains. 5G radio access technology, with all major band promotions, high digital streaming, and high reliability connections, is expected to carry a significant increase in the amount of data that requires network fraudsters to be expected in the coming years. The group used and adapted the appearance of the parameter of the virtual connector to the frequency bands, the 5G visual display contains a variety of wireless interface (AIVs), which can incorporate existing value technology, etc.

A) ventilate interface as mono-component. 5G surveillance supporting a wide range of applications and conditions including large-scale mobile broadband with 1000 enhanced capabilities, 99.999% reliability, and low ms overtime on a virtual wireless network, standard networks will be optimized for network performance and connectivity levels.

In this chapter, we introduce a number of RAN key mechanisms to ensure that they are configured in the context of the METIS II system, in particular interface management, robust vehicle control, network share management, and 5G network and LTE AIVs network [1] started by management interface so far the mobile network has been heavily influenced by writing. The 5G network creates initial problems with a disruptive pattern such as a seamless topological design process found on the same network i.e., using open-access nodes (or NN) open/closed. to detect network demand for continuous operation while managing a flexible interface simultaneously together up and down [2]. In addition, 5G RAN expects to control multiple groups (down and over 6GHz) and promotes 5G service with a variety of secondary requirements. In addition a powerful traffic control device is just as important as the unnecessary RAN power of a 5G network, which can be enhanced with strong definition and compelling Quality of Service (QoS). Traffic control is an important design for shared commitment and improved customer productivity [3]. Split navigation creates LTE traffic directory Many types of networks are investigated in [4], where all participants are related to a large cover that may be theirs service. Three network deployment and management allows network support to eliminate the RRM-related hypothesis [5] where multiplex pieces are assisted in sharing resources for this purpose, determined by the Agnostic Slice Enabler (AaSE) Awesome Air interface [6]. Loop for a Service Level Agreement (SLA) in accordance with AaSE having a network component. means the confirmation of the QoS order that AaSE contributes to the data processing dedicated to collecting SLA provision in bulk. state of the art embedding between different types of older AIV compatible with 3G and 4G is included in the standard power transfer of hard inter-RAT 0. large size with inter-RAT 0 hard drive e.g. in the case of 3G and 4G the longest delays and service are more disruptive as unreliable. The factory including high-end LTE is approved for its value in terms of very high reliability and strong bitrate in the form of 5G.

In this chapter, we introduce a number of key RAN approaches to ensure that they are prepared in the context of the METIS II system, particularly visual interface management, robust traffic control, network share management, and 5G and LTE AIVs internet performance [1] started with the management of the interface that is currently the most widely used mobile network.

7.1.1 Interface Management

The two suggested ways to deal with 5G high-density varies with the required flexibility based on advanced voice conversion systems but this does not mean that this will be a broad definition of part of 5G interference, which may require multiple approaches to avoid distortion and balance (Figures 7.1-7.3).

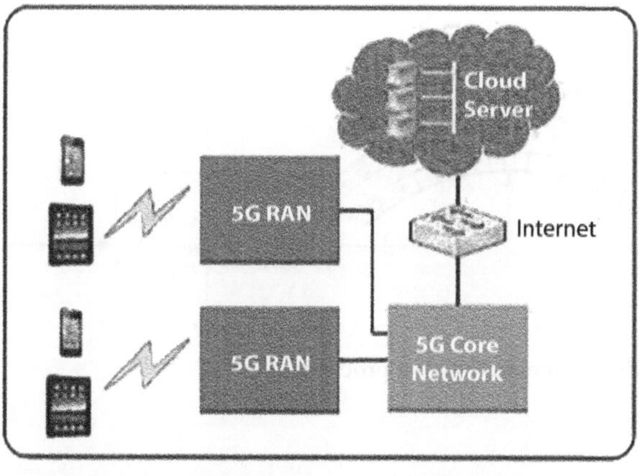

Figure 7.1 5G architecture RAN model.

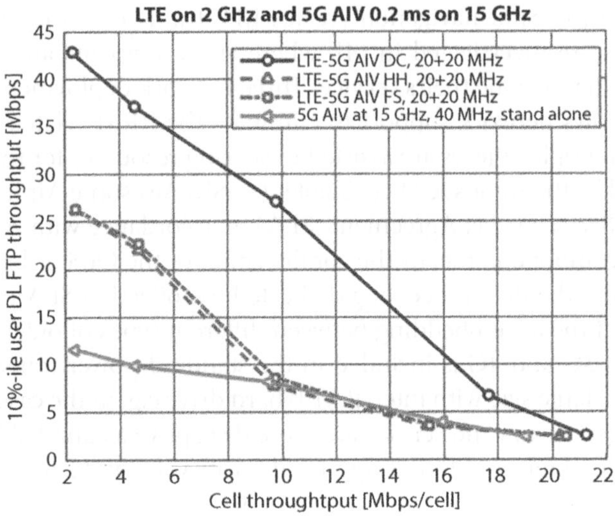

Figure 7.2 LTE graph.

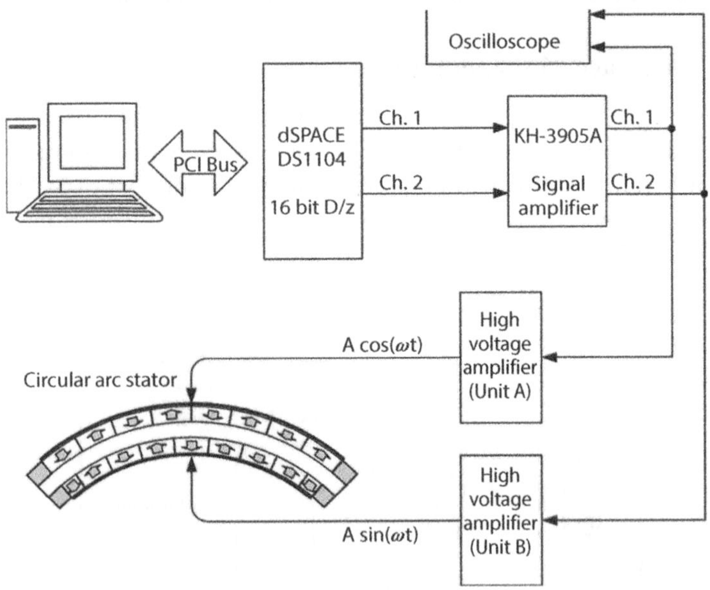

Figure 7.3 Hardware interface (signal processing).

7.1.2 Management of Conflict in Agile RM Framework

7.1.2.1 A Holistic View of Technology

The 5G RAN framework consists of a single active RAN framework. It is the effective foundation of the Interference Management block and seeks to develop an agile RM architecture [3]. The Agile RM framework includes a virtual air connection as well as comprehensive RM solutions.

Given the various objectives are prioritised to satisfy specified KPIs, some promising Disruption Management methods (Figure 7.4) are proposed in this context, with three main categories.

- Improving cell penetration by combining cracks or by creating "uninterrupted" areas to improve cell function.
- Toggling the power of micro cells enhances energy efficiency while also enabling coherent transmission modes (CoMP) techniques to deal with user downtime while also raising strain on nearby cells.
- Continued degradation by merging small cells and performing CoMP internal and external processes. This is explained in more detail later. The Figure 7.5 represents an overview

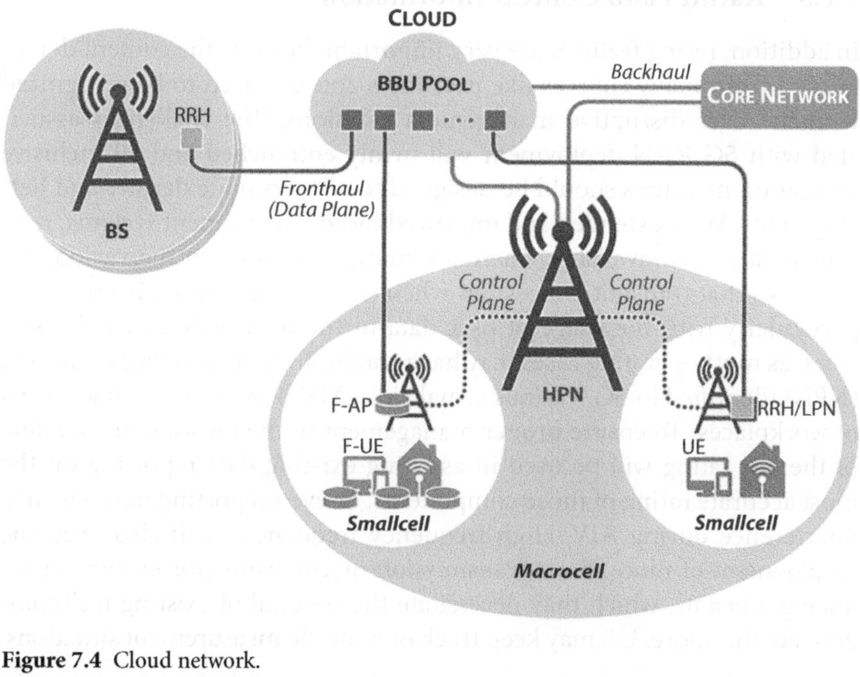

Figure 7.4 Cloud network.

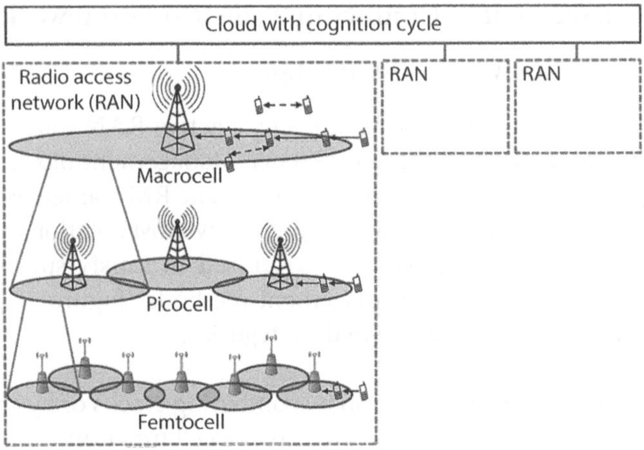

Figure 7.5 Cloud with cognition cycle.

of the proposed disruption control technology in the real Madrid grid, which may be thought of as a toolbox of solutions that can meet the needs of various operational and distribution scenarios, as explained in the following sections.

7.1.3 Ratings and Context Information

In addition, rating features are very important, because the content details needed from many sources like user tools and UE need to be recognized with the over disruptive management solutions. The challenges associated with 5G RAN deployment will firmly entrenched and all-inclusive mitigation measures should be designed to address node density and heterogeneity. With extended and improved radio management systems, real-time data is now available. There are numerous issues in the various 5G networks that transmit congestion when small cell cells are disseminated, particularly with the usage of user data in the distribution of radio services. as most operating cases may have conflicting performance indicators (KPIs) identified in 5G, various visual aids (AIVs) can be used in a variety of workplaces. To ensure proper management of the disorder, the context of the UE rating will be used in assisting existing data reporting on the most accurate rating of those components, or even reporting new data like interference during AIV. High-frequency frequencies will also need the employment of more direct transmission mechanisms (for example, producing a beam), which may necessitate the renewal of existing technologies. Furthermore, UE may keep track of multiple measurement situations,

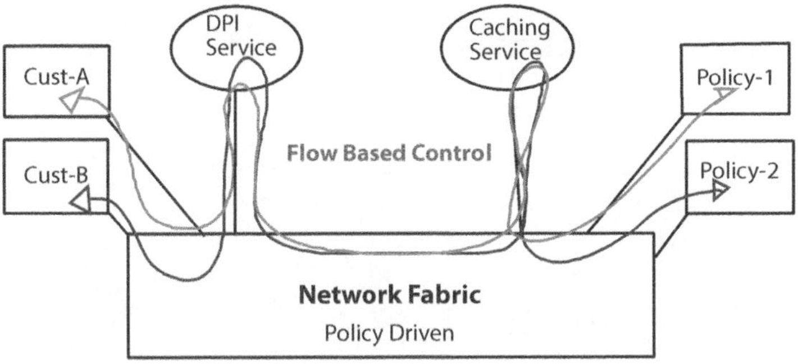

Figure 7.6 Network fabric.

such as AIV modifications (Figure 7.6). Furthermore, the varied 5G transmission conditions necessitate the development of appropriate UE-related measuring methods, which are equally essential elements (e.g., UE navigation mode, time adjustment/frequency). Extensions and updates in UE that are highlighted.

7.1.4 Dynamic Traffic Steering

7.1.4.1 Reduced-Overhead Interference Mitigation

Other 5G scenarios with poor backhaul presentation, such as the rapid deployment of tiny cells in low-income areas, have been observed previously. The improvement of the head connected with the disease control method is one of the primary challenges in generating a disruptive environment in EE in these circumstances.

In such instances, the development of disruptive, non-distributed distribution systems with low power and high latency (even below existing X2 interface values) will enable 5G implementation.

The method under investigation is based on the production of explicit (scattering and crushing) signals on complicated transmitted baseband signals. This increasing spread is done through consecutive transmission times (TTIs).

To implement this method to reduce interference associated to many access points all that is necessary is the sharing of information on orthogonal integration patterns to be used by each cell in the collection. The active blocks that are needed to apply the suggested reduction are shown. The orthogonal layout of the cell to which UE is attached should also be known to UE. It's straightforward to integrate it into the Down Link Control Channel because it's so high up. Figure 7.10: Functional Blocks for

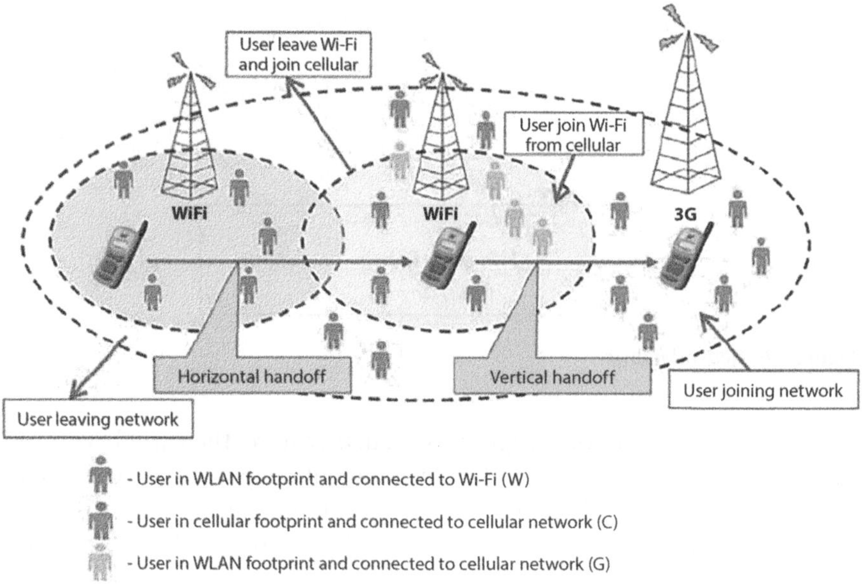

Figure 7.7 Wi-Fi architecture of cellular.

(Figure 7.7) Transmission and Reception. Once the pattern has been identified, UEs utilise ways to diminish and disperse to clear distortions caused by points of access to the same combination, since the emphasis introduces interaction between them.

It's worth noting that this interference migration process not only allows signals from various access points in the collection to communicate, yet, it moreover protects them from interference from other access points in the collection, improving SINR ratio in the same way that standard spectrum distribution methods do. In a frequency grid with a one-time frequency. Because numerous TTIs will be necessary to transfer any item to the UE designated for this approach to avoid interference, the adjustment will increase latency. However, not all EEs in this area are required to use them; only those who are currently experiencing problems as a result of cell disruption are required to do so.

7.1.5 Reducing Overhead Interference Mitigation

Other 5G scenarios with poor backhaul performance, such as the rapid deployment of tiny cells in low-income areas, have been observed previously. The improvement of the head connected with the disease control method is one of the primary challenges in generating a disruptive environment in EE in these circumstances.

In such instances, the development of disruptive, non-distributed distribution systems with low power and high latency (even below existing X2 interface values) will enable 5G implementation.

The method under investigation is based on the production of explicit (scattering and crushing) signals on complicated transmitted baseband signals. This spread widening occurs across a series of transmission times (TTIs).

To implement this method to reduce interference associated to many access points all that is necessary is the sharing of information on orthogonal integration patterns to be used by each cell in the collection. The active blocks that are needed to apply the suggested reduction are shown. The orthogonal layout of the cell to which UE is attached should also be known to UE. It's simple to incorporate into the Down Link Control Channel because it's so high up. Transmission and reception are depicted in Figure 7.10. Building Blocks concentrate your efforts to reduce high interference. Since the emphasis introduces interaction between them, UEs use strategies to diminish and scatter to remove distortions created by points of access to the same combination once the pattern is understood.

It's worth noting that this interference migration process not only allows signals from various access points in the collection to communicate, yet, it moreover protects them from interference from other access points in the collection, improving SINR ratio in the same way that standard spectrum distribution methods do in a frequency grid with a one-time frequency. Because numerous TTIs will be necessary to transfer any item to the UE designated for this approach to avoid interference, the adjustment will increase latency. However, not all EEs in this area are required to use them; only those who are currently experiencing problems as a result of cell disruption are required to do so.

7.1.5.1 RIM and 5G NR

The issue of remote interference caused by atmospheric imaging conditions has recently been explored in 3GPP 5G NR technical papers, for example, the main goal of this study was to look at possible mitigation programmes to enhance network stability against remote GNB interference, with a focus on large TDD semi-ready cells-static DL/UL. If the attacker's gNB can be detected and identified in TDD-based long-term evolution (TD-LTE) networks, the monitoring time (GP) between DL and UL transmission would already be able to extended. As can be seen conceptually in the lower part of Figure 7.8, the distance between the aggressor gNB and the victim's gNB greatly increases the amount of UL signals implicated in the victim's gNB.

For example, it is assumed that there is a carrier space (SCS) of a 15 kHz map with a signal length of 66.7 s and a cyclic (CP) start of 4.7, the distribution delay is similar to the five OFDM signals within 100 km between the abuser and gNBs for victims. When using synchronized TDD macrocellites to separate the DL and UL radio power, the effect of this pulse system is expected to be much greater on 5G NR networks than on TD-LTE networks when a large number of basic channels are connected. Due to the large number of potential sources of disruption and various distribution delays, effective methods of remote distortion control over NR (NR-RIM) should be considered.

The primary purpose of NR-RIM is to detect remote disturbances, recognize disruptive gNBs, and calculate distribution delays between attacker and gNB attackers or gNB groups. Two possible formulas were introduced and tested in a prior paper (RIM-RS). The 5G NR (CSI-RS) data reference signal is used in one OFDM (1OS) RIM-RS design, while the LTE RIM-RS design is used in two OFDM (2OS) RIM-RS designs.

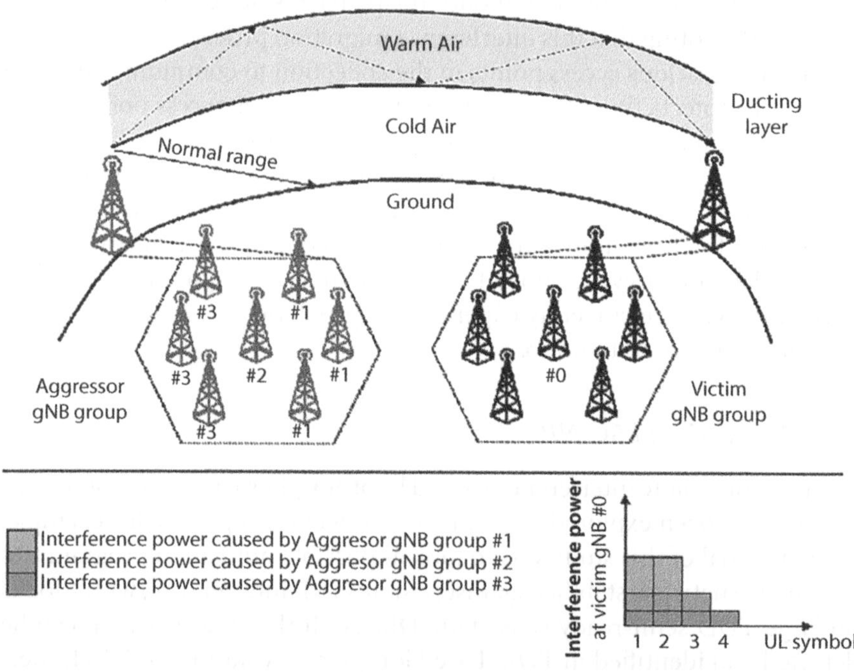

Figure 7.8 The problem of remote interference (RI) caused by the air ducting phenomenon is depicted in this diagram.

7.1.5.2 Contributions and Differences

This chapter builds on past work and findings based on a performance analysis of the RIM system, taking into account GNB's delayed distribution systems and RIM-RS transmission systems. Since landslides often affect large areas simultaneously. The network operator is expected to arrange gNBs in a specific region to send RIM-RS at the same time.

Advanced transmission schemes are studied in the proposed manner to determine how the available profits can be achieved when using common frequency tools to transfer all within this gNB phase, RIM-RS sequences or frequency multiplication changes to transfer the same or different base sequences are used. We also look at the efficiency gains that may be made by combining advanced acquisition techniques with a high number of acquisition windows, as well as the RIM-RS architecture. The short-term RIM-RS reduces a small increase in acquisition complexity, which is offset by the visible show results.

The results show that the improved receiver processing mechanism assures that both the RIM-RS 1OS and 2OS designs can accurately detect far distances beneath the space travel object. In RIM cases involving a large number of gNBs, gNB collections based on delayed use and use of the same frequency tools of all RIM-RS have been shown to provide excellent detection results. The techniques and results given here reveal fresh insights from the first scenario of RIM submission in the existence of a substantial number of gNBs that have yet to be published.

7.1.6 High-Level 5NR – RIM Framework

The RIM system uses OAM-based interventions to regulate and start or stop transmitting RIM-RS to TD-LTE networks (operation, management, and maintenance). As a result, there is much room for progress in terms of reducing manual interference in the event of problems while improving network performance and reliability. Depending on network design, the victim group or gNB of the victim located in the same location where the atmospheric bias occurs can transfer RIM-RS sequences at the same time as the ID group. Disruption can be assessed by the victim or on the victim's side as a result of channel retaliation. In this fashion, the original 3GPP NR-RIM data assumes that the DL and UL transmission parameters, as well as the reception parameters, are understood by the whole compatible macro cell network. Based on the examination of various processes, strong and flexible RIM schemes can be proposed in 5G NR. Network setups are used in static RIM schemes, which include setting up various monitoring times to accommodate large delays in supplying and transmitting grip strength. In addition, in the event of an RI, powerful TDD configurations such as the infusion of UL or DL on the sides of the victim or attacker, as well as the use of orthogonal frequency multiplexed services between aggressor DL and victim UL, may be used. However, there is a trade-off between the potential for operational performance and these RI reduction plans. On the other hand, at the expense of the victim, the lower antenna height or the adjustment of the antenna slope to reduce interference power can be used for network consumption and efficiency of the scale.

Flexible systems, such as using bean processing to reduce the signal received on the victim's side in terms of remote interference, can only be used in the event of remote interference to improve network performance. It is suggested, for example, that the beam removal work is done in an RI manner. The disruption rate at the horns pointing at the RI source may be 10 to 30 dB higher than at the horns pointing at other indicators due to the highly directed nature of the hot road demolition.

RIM-RS frameworks, as well as RIM, back-to-back signature network 5G network, are all possibilities. Figure 7.9 depicts a representation of the NR-RIM conceptual framework as it appears in 3GPP agreements. The research focuses on the dispersion of gNB clusters with distributed delay delays, in which gNBs in the same region might send the same they can be identified from other aggressor gNB groupings based on the time delay or sequence -RIM-RS.

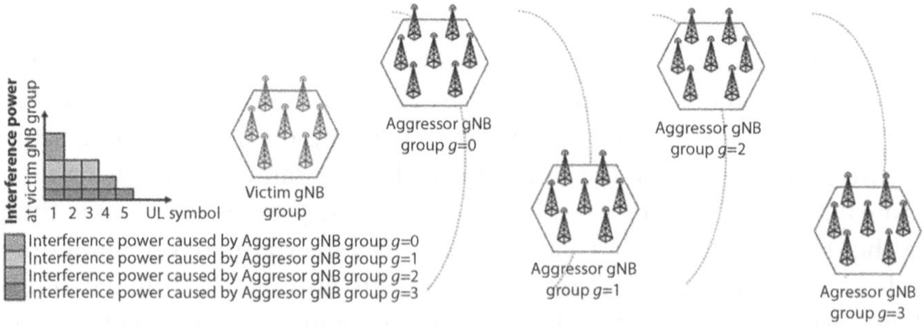

Figure 7.9 Signals were acquired from a number of remote aggressor gnb groups in various locations hit by air ducting phenomena, resulting in a sloping interference-over-thermal noise power response on the victim side.

7.1.7 OS RIM-RS

The actual NR reference signals can be configured in a variety of ways, thus they can be used as a starting point for building the RIM-RS. The available synchronisation signals for distant interference detection, such as main and auxiliary synchronisation signals may not be visible in a user resource (UE) piece attempting the first access procedures, therefore it is not regarded NR-RIM. The demodulation reference signal (DM-RS) is not considered for NRmRIM-RS production in 5G NR Release-15 since it is designed for channel restrictions in synchronised performance and has never been communicated to the local DL signal.

The NR CSI-RS, on the other hand, is unique to DL and can be mapped to the OFDM symbol within the space and shown to be transparent to UE, which is important when considering backward interactions. The CSI-RS structure provides for robust multi-resource allocation and duplicate generation in a temporal domain within a single OFDM signal thanks to a map-like resource base of the common domain.

The 1OS RIM-design, RS's which is based on 5G NR CSI-RS, allows a time zone duplication to be formed within a single OFDM transmission. The so-called camp-like representation can be received directly from the common domain in the RIM-RS projects, where the RS is broadcast to the whole Nrepth carriage, and equivalent to a smart domain duplicate form in a time domain within a single OFDM symbol. CSI-RS supports comb-type filtering in 5G NR Release-15, where the signal is distributed to all four-loaders in a common domain (called comb-4), and is then considered a four-fold duplicate structure at a time zone within a single signal of OFDM.

The CSI-RS power improvement in this scenario fixes the CSI-RS connected together the distributed physical downlink channel (PDSCH) for the PDSCH power related power boost (dB). As a result, RIM-RS is moved away from the shared user data channel.

7.1.7.1 OS RIM-RS

Following the RS requirements outlined in the 3GPP data, two other RIM-RS alternatives are being studied. RIM-RS uses a duplicate structure in TD-LTE networks so that identification is available in the OFDM signal segment in the receiver. In the 5G NR RIM test, signal detection was also considered to reduce the severity of our gnb deficiency.

RIM-RS uses repetitive configurations for TD-LTE networks, such as the first arbitrary entry structure (PRACH) for layouts 2 and 3. That the entire allocation bandwidth will be used. Similar to the LTE RIMRS, the 2OS RIM-RS is based on two identical copies of the whole PRACH series, with the CP length doubling to match the number of transmitted samples with two CP-OFDM markers. 2OS RIM-RS uses a layout similar to camp integration in a common domain, similar to the 1OS structure.

Furthermore, all cyclic shifts (CSs), or shifting the comb pattern to the frequency range, increase the detection efficiency. As a result, as shown in Figure 7.10, both RIM-RS designs, 1OS and 2OS, support two CS-comb-2 and four-comb-4 CS, respectively.

These RIM-RS frameworks are mainly aimed to increase acquisition efficiency in the RS-RIM system. The 1OS RIM-RS has a unique design, on the other hand, can be used to measure channel interference levels in nearby UE in the same region. To achieve this, new transfer strategies based on RIM-RS sequence information transferred and shared resources can be sent to gNB groups located in a particular area contrived by the seizure. As a result, surrounding gNB RIM-RS resources can be regarded non-zero power CSI-RS (NZP CSI-RS) or CSI-RS (ZP CSI-RS) power resources for CSI-RS (NZP CSI-RS). Because it interferes with the alignment of the 5G NR line signal, the 2OS RIM-RS model does not comply with interference requirements in this instance.

7.1.8 Radio-Access Network

LTE RAN uses a flat structure with one type of node2 - in NodeB. ENodeB manages all radio-related activities in one or more cells. Realize that

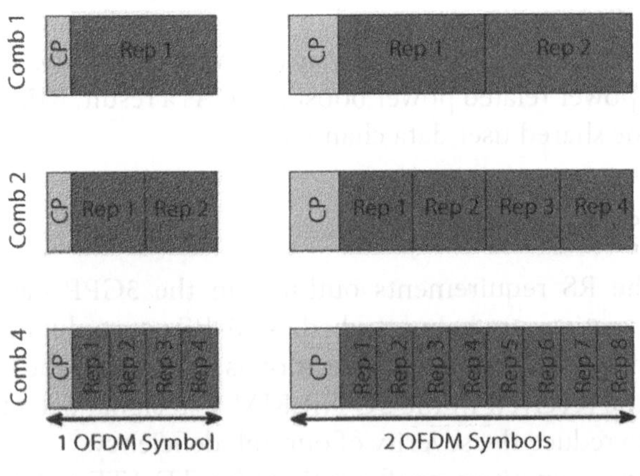

Figure 7.10 Time-domain resource allocation for the comb-1, comb-2, and comb-4 pilot patterns in 1os (left) and 2os (right) rim-rs designs.

eNodeB is a logical process and not a physical implementation. The most common use of eNodeB is a three-phase domain, where the main channel handles the transfer of three cells, even if other inputs are available, such as a single baseband processing unit to which remote radio heads are connected. One example is keeping a large number of internal cells in the same eNodeB, or fewer cells on the main route. As a result, while a simple channel can be used, it isn't the same as eNodeB.

ENodeB is connected to the EPC via the S1 interface, particularly S-GW via S1, S1-u, and MME user components via the S1 control component, S1-c. One eNodeB can be linked to multiple MMEs/S-GWs to share uploads and retrieval.

eNodeBs, which is connected by X2 interface, is widely used to support active mode navigation. This virtual interface can be used for multi-cell radio frequency management (RRM) operations such as cell disruption integration (ICIC). The X2 optical connector is used to support the non-slip movement between neighboring cells through packet transfer.

7.1.9 Improve RAN Coverage

RAN is a complex organization with various features. This is the case with the first release of 5G and a host of small cells connected to the simultaneous use of multiple radio access technology (RATs). These technologies provide power such as dual connectivity, spectrum power sharing, and 4G power reduction to 5G power outages. Aside from the explosion of basic channels and (Figure 7.11) cell sites due to 5G, coverage remains a problem for many providers.

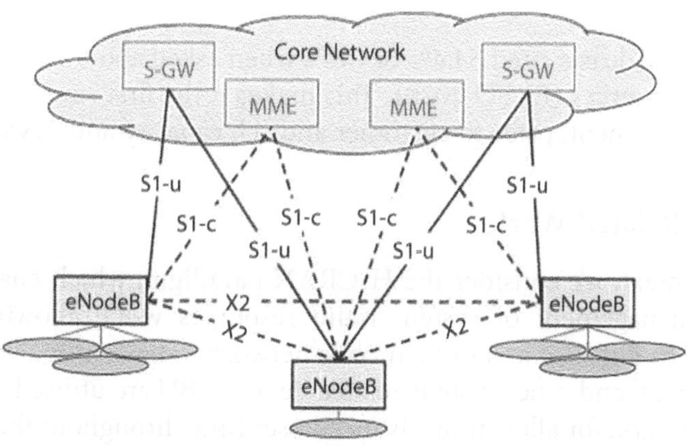

Figure 7.11 Radio communication systems.

7.1.9.1 Improving RAN Coverage Can Take Many Forms

- Install new cells to deal with spaces
- Switch the network to deal with hot spots
- Address black spots due to radio conditions and adjustments
- Improve RF vulnerability (cells with bad radio scores)
- Replace faulty network equipment

Nova RAN for EXFO simplifies the diagnosis and resolution of RAN coverage issues in a number of ways.

a) Collecting important performance data on RAN and spine is very easy. Nova RAN has large, straightforward color maps, with a resolution of up to 50m x 50m, showing real coverage, installation quality, highly efficient sites, and spaces. It can make geolocate calls to check the status of mobile radio and location and track the behavior of subscribers. It also detects and analyzes impact calls, by being able to detect the flow of calls and release agreements.
b) Getting started is easy and efficient. Nova RAN makes it possible to ensure that the correct configuration is present or that there are no inconsistencies in the setting. Before pressing a change of suspension, it mimics their impact with real-world subscriber details. Later, KPI styles confirm that the development was real.
c) Ensuring that repairs are quick and effective. Nova RAN uses real-world performance data to ensure that the configuration is enabled. This is similar to a 'visual' drive test.

Radio broadcasts cited 84.4% of CSPs when asked about their top three active CSP sources (EXFO 2020). This makes it the first source of performance impairment, prior to customer and cell capacity and devices.

7.1.10 Related Work

In this context, we consider the H-CRAN paradigm, which enables centralised management of system radio resources while allowing RRHs broadcast to different areas of a mobile network to be distinguished effectively. The UE and other system-related data in BBU are utilised to recommend a method for allocating advanced user time throughout the network in accordance with the ABS process.

As previously stated, the cloud computing capabilities of H-CRAN may be fully leveraged to minimise styles and improve spectrum efficiency. The topic has received a lot of attention in the literature, with a recent work on C-RAN and HetNet looking into the prospects of H-CRAN. H-CRAN-based methods, for example, are proposed in [4] for successfully regulating the amount of resources given to end users. This activity is a balancing act of fairness and inclusivity. Furthermore, [10] proposes the usage of new cloud cells in conjunction with HetNet in multi-congregation systems, with simulation results demonstrating the proposed infrastructure's efficiency in obtaining a higher system rating than a single band network. The authors of [11] propose a multi-resource sharing system based on the Indian bird game, which entails continuously monitoring C-RAN system circumstances and appropriately responding in a distributed manner. [6] also considers resource sharing, specifically spectrum sharing, in C-RAN-compatible systems. The research looked at the various frequency bands used by LTE and LTE-Advanced systems, with a particular emphasis on licenced shared access as a spectrum access solution. Furthermore, as discussed in [12], centralised C-RAN processing via Heterogonous Networks (HetNet) created new means of resolving internal disruptions by pooling resources across laptops, networks, and infrastructure, resulting in new ways of addressing internal disruptions. The benefits, limitations, and important technologies that allow for resource allocation on a C-RAN were explored at each of the levels above in the study. Furthermore, in [13], the authors developed a novel radio sharing system based on a collaborative game model that sensed the magnitude of global-based material and redefined accessible space. It is also required to address interference concerns between tiers in the HetNet environment, where different tiers share the spectrum, by lowering interference in the time zone, which is possible through the ABS process as investigated in [14, 15]. Building usage as a broader network use problem is studied in [14], while the authors presented a standard calculator that may be assessed by the network in [15].

One of the ways used in 3GPP to handle interruption is power control [16], while ABSF is another option to regulate intercell interference in time zones [17]. To lessen the amount of interruption experienced by nearby BSs, all BSs re-use the universal frequency and are reduced to a fraction of the framework. Our intended distraction programme is a sophisticated ABS system that allows several macro users to switch between empty frames deliberately for best performance. The concept of extensively sharing as a distraction-reduction strategy is not new; implementations (including a partial repetition system) have been proposed in the literature [18]. The cell is separated into an interior zone made up of users near BS and an external region made up of users near the cell's boundary in such systems. The waves used within the region might be shared by neighbouring BSs. There are two types of distraction management, according to HetNet: Multiband HetNet is the first, and it uses macro-BSs on the 2GHz band and pico-BSs on the 3GHz or 60GHz band. All BSs operate on the same frequency range in the second stage, disrupting intercell communication. In the present circumstance, dividing the spectrum into subbands to distribute high BS and low power is one of the proposed approaches in the literature to reduce distortion. Studies conducted in this context include, according to the concept of separating the spectrum from the given and assigned objects, for example, as [8]. A vote-sharing system for femtocells is proposed. It looks at practical issues such as hand delivery, coverage, and location distribution based on the computed area of disturbance determined by the co-channel femtocell area.

7.1.11 Conclusion

The concept proposed in this chapter describes some of the Random Access Network buttons that make some of the most salient features of the 5G Random Race Machine system that includes seamless management, static traffics control, powerful bit-slicing communication control, and 5-generation loose operation and loss transfer education.

References

1. B. Soret *et al.*, "Interference Coordination for Dense Wireless Networks", *IEEE Commun. Mag.*, vol. 53, no. 1, pp. 102-09, Jan. 2015.
2. K.I. Pedersen *et al.*, "Enhanced Inter-Cell Interference Coordination in Co-Channel Multi-Layer LTE-Advanced Networks", *IEEE Wireless Commun.*, vol. 20, no. 3, pp. 120-27, June 2013.
3. https://www.intechopen.com/books/broadband-communications-networks-recent-advances-and-lessons-from-practice/co-channel-interference-cancellation-for-5g-cellular-networks-deploying-radio-over-fiber-and-massive
4. ICT-317669 METIS project, "Final performance results and consolidated view on the most promising multi-node/multi-antenna transmission technologies," Deliverable D3.3, Feb. 2015.
5. 3GPP TR 36.814, "Evolved Universal Terrestrial Radio Access (E-UTRA); Further advancements for E-UTRA physical layer aspects (Release9)", V9.0.0, Mar. 2010.

6. S. Hong, M. Sagong, *et al.*, "Frequency and Quadrature-Amplitude Modulation for Downlink Cellular OFDMA Networks," IEEE J. Sel. Areas Commun., vol. 32, no. 6, pp.1256-1267, June 2014.
7. EARTH Project, see http://www.ict-earth.eu
8. 5GREEN Project, see http://wireless.kth.se/5green/
9. S. Zukang, *et al.*, "Dynamic uplink-downlink configuration and interference management in TD-LTE," IEEE Communications Magazine, vol.50, no.11, pp.51-59, November 2012.
10. R. Irmer, *et al.*, "Coordinated multipoint: Concepts, performance, and field trial results," IEEE Communications Magazine, vol.49, no.2, pp.102-111, February 2011.
11. ICT-317669 METIS project, "Scenarios, requirements, and KPIs for 5G mobile and wireless system", Deliverable D1.1, May 2013.
12. H. Celik and K. W. Sung, "On the Feasibility of Blind Dynamic TDD in Ultra-Dense Wireless Networks," in Vehicular Technology Conference (VTC Spring), 2015 IEEE 81st, vol., no., pp.1-5, 11-14 May 2015.
13. B. Debaillie, *et al.*, "A Flexible and Future-Proof Power Model for Cellular Base Stations," in Vehicular Technology Conference (VTC Spring), 2015 IEEE 81st, vol., no., pp.1-7, 11-14 May 2015.
14. C. Seol and K. Cheun, "A statistical inter-cell interference model for downlink cellular OFDMA networks under log-normal shadowing and multipath Rayleigh fading," IEEE Trans. Commun., vol. 57, no. 10, pp. 3069–3077, Oct. 2009.
15. I. Shomoronyand A. S. Avestimehr, "Worst-case additive noise in wireless networks," IEEE Trans. Inf. Theory, vol. 59, no. 6, pp. 3833–3847, Mar. 2013.
16. Brad A. Myers, "A Brief History of Human-Computer Interaction Technology", ACM,pp.44-54, 1998.
17. Brumby, D.P., Cutrell, E., Sarin, "The path to evaluating a new interface for the personal desktop search", Workshop paper presented at SIGCHI Conference on Human Factors in Computing Systems, R. Eye-tracking in practice CHI'06, Montreal, Canada, 2006.
18. Fawzi Al Sawalqa, "Customers Financial Needssatisfaction and Self-service Technology Banking The Case of Automatic Teller Machines (ATMs) in Jordan", International Journal of Business and Social Science, Vol. 3 No. 9, pp.191-200, 2012.
19. Curran, K. and King, "Investigating the Human-Computer Interaction Problems with Automated Teller Machine (ATM) Navigation Menus", Computer and Information Science, Vol. 1, no. 2, pp.34-51, 2008.
20. El-Haddad, A. B. and Almahmeed, M. A., "ATM banking behaviour in Kuwait: a consumer survey", International Journal of Bank Marketing, pp.25-32, 1992.
21. Marco Camilli and Massimiliano Dibitonto, "User-centered design approach for interactive kiosks: evaluation and redesign of an automaticteller machine:", Facing Complexity, pp. 85-91, 2010.

22. Di Nocera, F., Coletta, S., Ferlazzo, F., Renzi, P. "Numerosit del campione e stabili della stima di usability ottenuta mediante valutazioni soggettive.", Giornale Italianodi Psicologia, pp.695-704, 2009.
23. Nielsen, J., Perince, K. "EyetrackingWeb Usability. Berkeley, CA", NewRiders, 2010.
24. CharlesKalmanek,"ARetrospectiveViewofATM", ACM, 2002.
25. J.P.Kohli, "Designing an application-oriented terminal", National Computer Conference,1975.
26. Pepermans, R., Verleye, G. Van Capellen S., "Wall- banking", innovativeness and computer attitudes: 25-40-year-old ATM-users on the spot.", Journal of Economic Psychology, 17, pp. 731-748, 1996.
27. Mead, S., Walker, N. Cabrera, E.F., "Training older adults to use automatic teller machines", Human Factors, 38, pp.425-433, 1996. Rogers, W. A., Cabrera, E. F., Walker, N., Gilbert,
28. D. K. Fisk, A. D., "A survey of automatic teller machine usage across the adult lifespan.", Human Factors, 38, pp.156-166, 1996.
29. Rogers, W.A. Fisk, A.D., "ATM design andtraining issues", Ergonomics in Design, pp. 4-9, 1997.
30. Rogers, W.A., Gilbert, D.K., Cabrera, E.F., "An analysis of automatic teller machine usage by older adults: A structured interview approach", Applied Ergonomics, 3(28), pp.173-180, 1997.

8

Interference Techniques Based on Deep Learning in Wireless Networks

S. Arunprasath[1], A. Suresh[1*] and Jayden Khakurel[2]

[1]Department of Networking and Communications, School of Computing, College of Engineering and Technology, SRM Institute of Science and Technology, SRM Nagar, Kattankulathur, Chengalpattu, Chennai, Tamil Nadu, India
[2]Department of Child Psychiatry, University of Turku, Turku, Finland

Abstract
On small-scale wireless networks, traditional architectures deliver fast execution and nearly ideal performance, but when user density is high, the performance suffers significantly. The highly congested frequency band is one of the major problems for wireless communication technology. Numerous wireless users to share the same services in terms of time and/or frequency because of the constrained radio band. Interference develops from this configuration. The rising need for high data-rate transmission is driving current development in wireless communication technology. This has led to a major improvement in wireless networks' performance in recent years. However, there is still a pressing demand for more effective wireless communications given the widespread usage of smart phones, laptops, and multimedia devices. For numerous networks, including cellular networks, wireless lan, and wireless ad-hoc networks, communication inside the presence significant interference has indeed been taken into consideration. One of the finest techniques for reducing wireless communication interference is deep learning. Each one operates by taking lessons from a dataset. One of the cutting-edge technologies that has been applied to address current wireless communication problems is deep learning. Deep learning employs two distinct methods: data-driven approach and model-driven approach. Neural networks were used in place of the traditional building blocks in the data-driven approach. The second method is the model-driven approach, which substitutes a neural network for parts of the traditional algorithms' strategies.

**Corresponding author*: prisu6esh@yahoo.com; suresha2@srmist.edu.in

A. Suresh, J. Ramkumar, M. Baskar and Ali Kashif Bashir (eds.) Resource Management in Advanced Wireless Networks, (161–182) © 2023 Scrivener Publishing LLC

***Keywords*:** Wireless communication, deep learning multiple-input multiple-output (MIMO), graph neural networks, interference suppression, interference alignment, interference cancellation

8.1 Introduction

Wireless communication is increasingly being used. By the end of 2027, the traffic on wireless communications systems is expected to reach 368EB each month. System traffic is caused by the development of more augmented, virtual, and mixed reality applications. Video traffic is anticipated to reach 69% in 2027. Wireless communications systems will have to deal with a rising demand for faster data speeds in the future.

Spectrum is costly and efficiently managed. The same frequency resource must be shared by multiple users for a wireless network to be effective. It might obstruct things. One of the main issues limiting the effectiveness of wireless communication is interference. Interference is excess noise that makes it difficult to clearly receive wireless communications. Interference Alignment (IA) multiple transmitter-receiver pairs can use the same radio spectrum resources when they are aligned. It is a method of pre coding that lines up transmission signals. Prior to the invention of interference alignment, interference-free communication was achieved via TDMA techniques. To completely align the interference, interference alignment require accurate channel state information.

Interference Alignment is based on several idealised hypotheses that are unfeasible for real-world communication systems. Because of this, before IA is implemented in the wireless communication systems of today, all these presumptions must be adjusted and worst-case scenarios must be taken into account. Coding is used to align all interference signals into the smaller region in order to accomplish IA. Because of this, if there are more interference signals, require more dimensions in order to align them. Prior studies have demonstrated that the required signalling dimensions for frequency domain alignment increase faster than exponentially as the number of user increases. In the situation of Multiple Input Multiple Output (MIMO) IA, in which more users may be handled so as the number of antennas increases linearly with network size, this dimension issue is significantly less severe. In light of the findings of earlier research, MIMO systems are more likely to integrate IA. In the sphere of conventional communication, MIMO technology is frequently employed to enhance communication quality [1]. To accomplish interference alignment, pre-coders are used. In order to ensure accurate Channel State Information (CSI), enough resources must be devoted to pilot transmissions and CSI feedback. The performance of IA can be severely limited by high-mobility fading technologies, which is why channel modifications must be implemented as rapidly as feasible. IA pre coders heavily rely on precise information of CSI.

Due to the growing demand for more signalled dimensions, the skyrocketing cost of CSI collection, and synchronisation, applying IA techniques to large-scale systems like cellular networks is incredibly challenging. Notwithstanding its complexity, it can be made simpler because not all interfering linkages should be treated equally as not every interfering links are significant. With encouraging outcomes, deep learning-based methods have been created to address difficult wireless communications issues. Deep learning approaches have already been shown to significantly increase interference suppression compared to traditional suppression methods. Additionally, deep learning approaches are helpful in learning complicated non-linear selection limits because they can recognise complex non-linear relationships in the input data [2]. High levels of noise can be effectively reduced by careful management of signal processing methods, increasing the SNR of both time-frequency and evoked magnetic field analyses at the sensor and source levels [3]. Initial attempts used neural network topologies that were adapted from programmes like computer vision. Resource allocation issues that cannot be addressed or even modelled by the conventional optimization framework can be solved using deep learning and deep reinforcement learning, respectively [4]. Convolutional and LSTM layers combined with an auto encoder architecture are particularly efficient. The practicality of implementations in next-generation communication systems is demonstrated by estimates of power and latency consumption [5]. They frequently produce poor results in large-scale networks and demand a lot of training data. Recently, these concerns have been addressed using graph neural networks (GNNs), which may effectively exploit the domain knowledge or the graph structure for wireless communication problems. Although GNN-based approaches can adapt well to different system settings and achieve near-optimal performance in large scale networks, their theoretical underpinnings and design guidelines are yet unclear, which may make it challenging to gain experience.

Compared to conventional neural architectures, GNNs function almost similarly in wireless networks with quite a significantly smaller training sample size.

For a design framework that can be used to address common design difficulties in wireless networks, including graph modelling, the neural design phase, including theory-guided performance enhancement. To legitimize the usage of GNNs for addressing wireless communication difficulties, provide an unified framework that integrates optimization-based methods and GNN-based algorithms.

8.2 Literature Review

Deep learning algorithms are useful for comprehending complex non-linear decision limitations because they can learn intricate non-linear connections in the input data. With hopeful outcomes, these strategies have been created to address severe wireless communications challenges. Initially, neural network designs that were taken from computer vision applications were adopted. They frequently produce poor results in large-scale networks and demand a lot of training data. Graph neural networks (GNNs) has lately been used to address these challenges because they can successfully utilise the domain knowledge, or the graph structure in wireless communication difficulties. The spatial integration in CNNs is expanded to graphs by GNNs. A CNN layer's individual hidden states are maintained by each pixel by convolution with nearby pixels. Because the signal is conveyed from the upper layer through to the highest pool layer, CNN has the benefit of translation invariance that enables the converted weights to detect patterns at any place in any layer. This capability allows the network to recognise patterns efficiently [6].

Similar to this, each node in one GNN layer changes its own hidden layer using data gathered from its neighbouring nodes. Although GNN-based methods can attain near-optimal efficacy in complex systems and generalise well in a variety of system situations, their theoretical foundations and design principles are yet unknown, which could make it difficult to put them into practise. Compared to conventional neural architectures, GNNs function almost as well in wireless networks with a significantly smaller training sample size. Selecting the GNN architecture that performs well on the related graph is an easy approach to choose from the variety of GNN designs that the machine learning profession has recently developed for various types of graphs. The framework drastically lowers overheads while making cloud data exchange simpler, secure, and more visible [7]. Add end-to-end training on atop of that to achieve superior performance to explicit approaches [8]. The generalisation error and amount of training samples needed for GNNs to tackle an optimization model on an n-node graph, respectively, are O (n) and O (n^2) times lower than those of unstructured multi-layer senses. Situations that pose a hazard to human life come from significant interference on wireless networks from various technology and devices [9]. Create GNN designs that make learning $h^{(t)}$, $g^{(t)}{}_{1}$, and $g^{(t)}{}_2$ simpler in order to further increase the bound in, or to lessen wireless communication problems. There are four widely used strategies, which are listed below.

i. To simplify the message passing scheme
ii. To find a suitable pooling function
iii. To improve the aggregation neural network
iv. To apply input embedding

The complexity of the procedures typically rises as a consequence of the re-insertion, which depends on the volume of new messages received by a node, in order to **streamline the message passing scheme**. Therefore, by making the message transfer mechanism simpler, the intricacy can be decreased.

To find a suitable pooling function-Max, Sum, and Focus Aggregation are the most often used pooling functions. The maximum aggregation concentrates on a few of the most significant neighbours. It is appropriate when the variables are noisy or the neighbour's influence is weak. Summaries of the neighbours' statistics are provided via the sum aggregation. When trying to compile a list of the neighbours, it is helpful. The neighbours that have a strong link with the current node are the focus of the attention-based aggregation. When the nodes' mutual correlations are crucial, it can be used.

To improve the aggregation neural network-Combined neural network of types q1 and q2. The q1 and q2 architectures of most GNNs are one-layer MLPs. These neural networks can be created using conventional unrolling approaches when a nice classic DMP method is available.

To apply input embedding-The neural network frequently disregards the edge feature when its dimension is significantly lower than the hidden dimension. It is typically advantageous in this circumstance to use an embedding neural net that lifts the edge characteristic into a multidimensional space. There are numerous methods for managing interference, including interference exploitation, interference neutralisation, interference alignment, and interference neutralisation. It is necessary for effective interference management that there be a distinction between helpful signals and interfering data in some areas. While advanced interference management strategies should be developed to increase system performance, interference management is essential to achieving the ideal balance between the usable signals and interference signals [10].

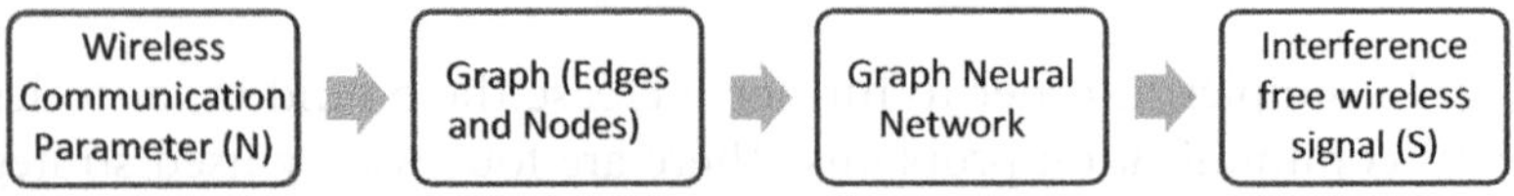

Figure 8.1 Graph neural networks method.

In Figure 8.1 graph neural networks method Graph parameters for a wireless network are node and edge. The GNN architecture is designed using the graph type. Adjusted neuronal architecture to increase effectiveness. GNN model final output interference-free wireless signal. 2 types of GNN layer are Aggregation and Combination:

Aggregation:
The outputs from the previous layer are combined by the n-th node's neighbours using a neural network, and then a pooling function is used. The update can be expressed as:

$$N_{(t)}\, n = KLj{\in}N\,(n)\; b_{(t)}\, 1\, (d_{(t-1)}\, n,\, d_{(t-1)}\, j,\, A\, (j, n, :)\; |M_{(t)}\, 1),$$

where,

N (n) is the set of neighbours of node n,

KLj∈N (n) is a pooling function, (Pooling function reduce the spatial size of the input)

$b_{(t)}$ 1 is the aggregation neural network at the t-th layer,

$M_{(t)}$ 1 denotes the learnable weights of this neural network.

Combination:
A secondary neural network is utilised to analyse the data and generate output at each node after the information has been aggregated. In the manner outlined below, the collected information is precisely combined with the node's unique data.

$$d_{(t)}k = c_{(t)}{}^2\, (d_{(t-1)}\, k,\, Z(k,:),\, N_{(t)}\, n \,|\, W_{(t)}{}^2\,),$$

In this equation, the learnable modeling variables W (t) 2 and the combined neural network $c_{(t)}$ 2 are represented. Designing the pooling function is necessary before using a GNN on a specific graph. Deep Learning (DL) has immensely benefitted future wireless communications (FWC). In which deep reinforcement learning, deep unsupervised, and deep supervised learning are all included. A wide variety of issues, including prediction, optimization, and decision-making in FWCs, can be resolved using DL approaches. Without quick and easy access to mathematical formulations, it would be more difficult to design and implement wireless communications capabilities, especially in cases when mathematical models are unable to adequately define the problems under consideration. Yet, DL for wireless technology is still in adolescence, and its advantages over a very well mathematical models have not yet been thoroughly examined and substantiated. A variety of other difficulties, including as communication system, model construction, pattern discovery, model deployment, and cyber security, are also brought up by the employment of DL techniques in wireless communications and must be carefully addressed.

Furthermore, the recently developed FWC situations, such as Extreme Highly strung Reliable Low Latency Communications networks, Massive Machine-to-machine Communications, Remote Controlled Vehicles, Industrial IoT, and combined Communication and Detecting, have created new challenges and research activities for the design and optimization of a functional units in FWCs, as well as how to create DL-based solutions to address the demanding performance standards. Due to the limited of wireless spectrum accessible and the increasing demand for frequency utilisation brought on by recent developments in wireless communication technology, interference is still an issue.

Despite recent improvements in the treatment of interference problems, interference still poses a major obstacle to efficient spectrum use. This is partially because more people are using managed, license-free shared bands and other opportunistic spectrum allocation techniques. The issue of interference has consistently dominated and plagued the wireless communications industry.

Recent technological advancements such as 5G communications technology, the internet of things, long-term advancement non-licensed, LTE licenced assisted availability, Citizen Broadband Wireless Provider spectral response, vehicle-to-everything, and the growing number of fairly priced application radios recommend that interference could still be a issue.

The interference from homogenous and heterogeneous technologies will continue to have an impact on reliable transmission due to the sheer number of wireless networks using the frequency band and the anticipated expansion of huge machine-type communication. For instance, when valid secondary users share a spectrum band, there may be interference between them; however, rogue secondary users (SUs) may also purposefully interfere with both legitimate secondary users and legacy users.

Wi-Fi unlicensed applications have been hampered by real-time monitoring software, Bluetooth connectivity, smart homes, and, more recently, the emergence of an unrestricted mobile network in these bands. Others have used PHY layer approaches to reduce interference, while early interference suppression strategies used interference minimization by splitting signals inside the time, frequency, code, or space domains. A wide variety of interference usage could present difficult barriers. A few examples of adaptive spectrum utilisation are automotive sensors, non-orthogonal channel allocation systems, the ubiquitous internet of things, full duplex frameworks, and the coexistence of different technologies.

Co-channel interference & adjacent channel interference are the two basic types of interference in a communication channel. The primary distinction between the two, despite the fact that they both take place within the same frequency band, is that the former occurs inside two distinct frequency bands. Homogeneous technique interference and heterogeneous technique interference are the two categories into which technology-related interference could be separated. While the latter suggests interference between separate technologies, the former refers to signals from the same medium interfering with one another. Additionally, when both transmission and reception happen simultaneously in a frequency region, full-duplex systems designed to increase spectral efficiency encounter self-interference (SI). Channel Gain, or Single-Noise-Ratio (SNR) in wireless communication, is calculated based on the Noisy Wireless Communication Signal (NWCS).Where there is noisy wireless communication Signals are combined in the form of noisy and clean signals.

$$\mathbf{NWCS\ (n) = CS\ (n) + NS\ (n)}$$

NWCS (n) - Noisy Wireless communication Signal
CS (n) - Clean Signal
NS (n) - Noise Signal

Wireless communication frame length of 'len' and 'k' point discrete Fourier transform,

$$\mathbf{NWCS_{len}\ (k) = CS_{len}\ (k) + NS_{len}(k) \quad k=\{0,1,2,\ldots,k\text{-}1\}}$$

For frequency bins, the gain function to calculate the estimated clean wireless signal, use Gain_{len} (k).

$$\mathbf{CS_{len}\ (k) = Gain_{len}\ (k) * NWCS_{len}\ (k)}$$

Gain function depends on the Prior SNR and Post SNR

$$\mathbf{Gain}_{\mathbf{len}}\ \mathbf{(k) = g\ (Prior\ SNR}_{\mathbf{len}}\ ,\ \mathbf{Post\ SNR}_{\mathbf{len}}\mathbf{)}$$

Here, g(.) is a commonly used minimal mean square error log spectrum amplitude [11].

8.3 Suppression Techniques

Conventional interference suppression solutions use the interference categories to determine the appropriate line of action. PHY layer systems using blind source separation (BSS) methods are a few of these methods.In order to deal with interference at the receiver, PHY layer approaches focus on adding redundancy or improving PHY layer design. In order to reduce interference, they look at the best PHY layer designs using tried-and-true methodologies like heuristic models, optimization methods, and greedy algorithms.

Figure 8.2 shows the LSTM-based suppression technique. This setup combines a feed-forward layer (FF) and an LSTM. There are 425 nodes that are used in this suppression technique. The Relu activation function is used. Spectral redundancies in orthogonal frequencies division multiplexed signals is utilised for interference reduction methods which use time-varying frequency filters known as TV-FRESH filters to lessen linear regularity modified radar interference. A signal can also be cleared of interference using an adaptive filter without losing important data. The capacity of adaptive filters to modify the filtering criteria according to the interference type that needs to be suppressed is one of its important features. In earlier investigations, interfering signals were identified and suppressed using signal attributes such inherent cyclostationary features, signal envelope, and channel bandwidth.

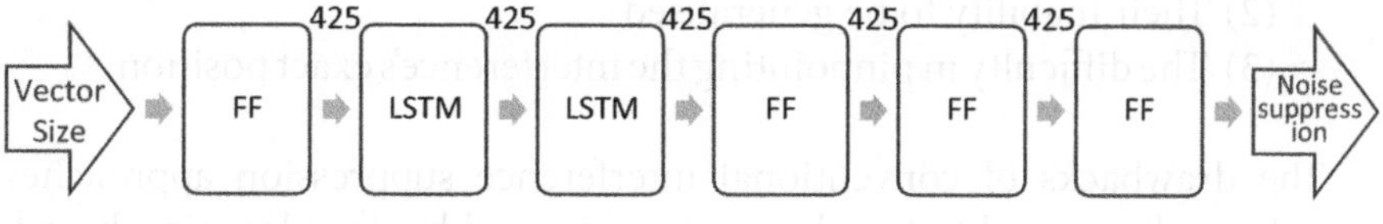

Figure 8.2 Suppression techniques.

Time-varying interference produces time-varying weights that have the propensity to reduce the efficacy with conventional adaptive filters such Minimum Mean Squares. This is the fundamental issue with PHY layer approaches that use predicted time analytics to adjust the filter weights. Furthermore, despite the variety and abundance of wireless devices that can be found in modern wireless environments, a lack of scalability may come from the dependency on domain knowledge or specific signal information. As part of their investigation into BSS approaches, they created a method to forecast interfering signals that are available at the same frequency as the desired signal by applying a rough prediction of multi-path propagation algorithm. A group of signals is distinguished in BSS from the set of mixed messages with or without information of the mixing process or the source signals. Simply said, BSS is used to isolate the main component of a differential amplifier. The noise-free portion of a broadcast that the receiver intends to study or retrieve is referred to as the "signal-of-interest" (SOI).

Although much of the research in the literature focused on decoding audio signals with lower dimensionality, BSS algorithms have also been used to multi-dimensional data. Principal component analysis (PCA), for instance, is a BSS method that has been used to the dimensionality reduction problem in the image processing area. They used the predicted waveform to reconstruct interference elements at the receiver for their BSS approach, and they then removed these components from of the time-domain data. Using a time-domain method to distinguish the signal of interest (SOI) from the interference signal of the radar system. They next reconstructed the signal while also subtracting the approximation interference signal from of the original signal by first estimating the signal parameters using a moment window that contained just those discrete Fourier transforms. These approaches' fundamental drawback is that they typically call for extremely precise interference position, duration, and other characteristic identification. Three major limitations of conventional interference suppression techniques.

(1) Relying on interference properties and domain knowledge;
(2) Their inability to be generalised
(3) The difficulty in pinpointing the interference's exact position.

The drawbacks of conventional interference suppression approaches have been discovered to be adequately addressed by deep learning-based, model-free interference suppression strategies. While real-time interference modelling, for example, could be difficult, a model-free approach might make them easier to deploy in the actual world and require less domain knowledge. It could appear extremely challenging to accomplish this goal given the highly stochastic nature of a wireless radio environment, particularly in the situation when failure to suppress such occurrences could have a severe influence on performance.

Deep learning techniques might be able to learn from the data without needing any prior knowledge to minimise interference. Although it is difficult to generate enough datasets for different channel environment realisations, software-defined radios (SDRs), that are affordable which can be used to produce data in several of channel situations, have arranged the data collection process. Deep learning frequently needs access to a vast data store. It's important to understand that, even if it might take a bit of time, learning for interference cancellation will eventually be helpful to the wireless transmission research group as long as it is explored and used. Deep learning's capacity to build strong, non-linear interconnections for interference suppression, which would be challenging to build with conventional suppression techniques, is another important advantage. When the border is complex and non-linear, many traditional interference suppression techniques are unreliable. The ability to learn complex non-linear decision boundaries is a major factor in the interference suppression group's use of deep learning techniques. Deep learning can help resolve the scalability issue, which is a challenge for many traditional suppression strategies.

8.3.1 Deep Learning Based Approach

To execute the goal of improving wireless communication, a deep learning-based technique uses neural network models that have been trained beforehand using a collection of learned data.

$$\mathbf{WS}_{\mathbf{Enhance}} = \mathbf{f}\,(\mathbf{V}_{\mathbf{Output}}, \mathbf{H}_{\mathbf{len\text{-}1}}, \mathbf{P}_{\mathbf{training}})$$

$WS_{Enhance}$ - Wireless communication Signal Enhancement
$f(.)$ - Nonlinear composite function
$P_{training}$ - Trainable parameter
$H_{len\text{-}1}$ - Hidden state
V_{Output} - Output vector

Deep learning approaches can be used to address the shortcomings of interference abandonment solutions based on deep learning.

The unreliability of the wireless link, difficulties with open set identification, and a lack of scaling are a few of these limitations. Deep learning techniques outperform non-deep learning methods in a variety of channel situations while also reducing computing complexity compared to non-learning suppression methods.

8.3.2 Adjacent Channel Interference (ACI)

When a signal's power interferes with another communication in a different channel, it is referred to as adjacent channel interference. ACI can happen for a number of causes, such as insufficient filtering in frequency modulated (FM) systems. ACI can also result from poor frequency regulation in either the reference channel or the interfering channel. ACI can sometimes be avoided using interference avoidance strategies. Despite the fact that there are a total of 11 channels accessible on the 2.4 GHz Wi-Fi band, it is advised that US Wi-Fi users use non-overlapping channels.

8.3.3 Co-Channel Interference

On the other hand, co-channel interference (CCI) happens when two equipment or communicators start to interfere with one another while using the same transmission channel. When frequency bands are reused in mobile communications applications, CCI is much more evident. To decrease CCI, interference avoidance techniques alone are insufficient. Concentrate on the more challenging CCI problem in this survey. How the SOI is hampered by CCI and ACI. It's also vital to remember that the receiver is where the receiver performs the data gathering, offline learning, and other deep learning components. An example of how A.I. systems or deep learning can be used to minimise or reduce interference.

8.3.4 Self-Interference

Self-interference (SI) is a reasonably typical phenomenon in full-duplex networks, when receivers and transmitters use the same frequency range for simultaneous communication. Since it improves spectral efficiency while better utilising the available spectrum, in-band full-duplex transmission & receive at the same frequency at the same time is very advantageous for cognitive radio (CR) broadcast systems.Research has been done on several interference cancellation techniques, such as aerial, analogue & digital cancellations.

8.3.5 Homogeneous Technology Interference

When numerous signals emanating from the same wireless technology interfere with each other, this is referred to as homogeneous technological interference, commonly known as the same interference. They can show up on two distinct channels or just one. There may every once in a while be more than a interfering signal, depending on the type of frequency range the dominant message is using.

The usage of different access protocols like IEEE 802.15.1 & IEEE 802.15.4 is typically what causes this interference. Adaptive methods, graph colours, TDMA-based approaches, as well as met heuristic, are some examples of interference suppression strategies. Homogeneous technological interference avoidance solutions for wireless body area networks (WBAN) include the CSMA/CA-based method. Additionally susceptible to disruption from homogeneous technology is the multiplexing method for cellular communications known as code division multiple-access (CDMA). Other communication methods and spread spectrum systems use the same band as CDMA. Each users in a widespread CDMA system has a distinct electronic pseudo-noise (PN) code which aids the receiver in differentiating them from other members of the system. In order to correctly accomplish this. To reduce homogeneous technical interference in CDMA systems throughout the past three decades, a number of interference suppression approaches, particularly linear and non-linear estimating techniques, have been developed. As a result, several DL-based interference reduction systems studied more heterogeneous technological interference than homogeneous technological interference.

8.3.6 Heterogeneous Technology Interference

It is essential to comprehend Cross technology interference (CTI), also known as heterogeneous technology interference (HTI), it is a multiple wireless technologies to coexist in unlicensed frequency bands. CTI happens when signals from other technologies interfere with a wireless technology. Typical comparisons include Wi-Fi radar3, Wi-Fi approved LTE, Wi-Fi uncontrolled LTE version 2, Wi-Fi, Wi-Fi Control Solution Communication (DSRC), and Wi-Fi. Studies have recommended a variety of interference reduction strategies in order to understand how these various technologies interact due to coexistence concerns that may cause interference. Interference-aware methods can take advantage of the distinctive, recognised patterns that CTI present. When IEEE 802.15.1 signals interfered with an IEEE 802.11 b/g signal, the impact of CTI was applied, and it was discovered that the IEEE 802.15.1 frequencies were comparatively benign in comparison to the opposite. This is because, due to their lower bandwidth, IEEE 802.15.4 & IEEE 802.15.1 frequencies exhibit less overlapping interference than IEEE 802.11 b/g signals. However, when using the same spectrum as other devices, CTI can have a very negative influence on low-power wireless devices. The 2.4 GHz band is shared by IEEE 802.11, Wireless, Bluetooth low energy (BLE), wireless USB, IEEE 802.15.4, 5, and other wireless technologies. Despite having different power levels & occupied spectrum widths, both technologies could communicate with one another because they are both sharing the congested unlicensed 2.4 GHz range.

8.4 Classification of Interference Suppression Techniques

The five various forms of interference suppression strategies can be categorised into the following groups: baseline, diversity-based, lower limit, interpolation-based, and DL based. Polynomial-based SI cancellation strategies have also been investigated by researchers. Full-duplex systems usually employ polynomial-based methods since they generate many parameters and make implementation challenging. Although the study's main focus is on deep learning-based methods, this section also contains a brief examination of non-deep learning alternatives to give the comparison some context. An interference suppression classification that encompasses both deep learning & non-deep learning methods is illustrated by the comparison. The Reconfigurable intelligent surface (RIS) approach is used for railway wireless communication to suppress interference [12]. Before using ML or deep learning techniques to do so, experts tried a variety of tactics to reduce or remove interference. Due to the huge number of limited wireless networks which were both power and computation constrained, it has been hypothesised that the PHY layer information combined temporal variation of the wireless channel will minimise heterogeneous technological interference. Corrupted frame values for the link quality indicator (LQI) are tracked, and traditional threshold procedures are used to find interference. The received signal strength indicator (RSSI) variations are gathered when interference is found. The received signal is sampled in order to record the RSSI variations. The detecting algorithm is then given the RSSI readings. The incorrect packet can be recreated through a subsequent interfering resend and the buffering of the properly decoded data.

Understand that earlier research has looked into interference avoidance strategies including transmit power control and dynamic frequency switching, among others, however the focus of our survey is on interference suppression methods. Zeroing interference sampling and IMAT time domain variable interpolation ramp filtering (RF min) are two examples of these techniques. The simplest way to suppress interference may be to zero out the interference components, but this could lower SNR or perhaps potentially blank out the signal itself. The range spectrum over ramps' diversity is taken advantage of by the RFmin approach. In IMAT methods, the corresponding negative delay is continuously patched by replacing the space with the fourier transforms of the real dominating spectral component. Interfering sampling is first neutralised. The use of a parameterized necessary to clearly to fill in gaps left by zeroing off interfering signals distinguishes the time domain parameter interpolation method from the IMAT approach.

It is worth noting that the collection of non-deep learning strategies for interference suppression offered in this part is just not exhaustive and only highlights some of those techniques because deep learning methods in interference cancellation are all in fact the main subject of this study. Readers who are curious in these additional methods can consult the most widely used methods for each category. Supervised learning & unsupervised learning are the two main categories under which deep learning techniques can be divided. Supplied learning methods, CNN-based suppression methods, DNN methods, and RNN methods are all included in method 6. Using a hybrid DNN model, non-convex optimization issues can be solved by approximating the near-optimal solution [13].

Both ML-based channel state information (CSI) methods and auto encoder-based suppression techniques have been investigated for determining the appropriate transmit power level to minimise interference. The channel state information required for coordination is frequently gathered by a computing node in a centralised environment [14]. This method heavily depends on a precise estimate of the transmit power needed in a wireless sensor network (WSN) situation to prevent interference. There are alternative methods for detecting interference from household appliances, Bluetooth, and Wi-Fi, such as employing light extraction of features and lightweight decision trees. Traditional interference suppression approaches could not be scaled to handle any resulting interference issues since there are so many devices vying for the same wireless band. Researchers have begun extending their evaluation and investigating their potential for interference suppression because deep learning approaches have been demonstrated to increase accuracy of classification for a variety of interference classification or identification.

8.5 A Blind Approach of Interference Cancellation Using Neural Networks

Neural networks can be used to solve the massive problem of interference in wideband cellular communication. During the training phase, the network makes use of training data to learn the intricate behaviour of wideband interference. During data transmission, the trained model subsequently functions as a real-time signal processing module. BNN achieves better accuracy by using back propagation [15]. Optimise the neural network for real-time operation using model compression approaches such as network pruning and coarse quantization to address the computation load and latency difficulties. The Multiple Interference scheme is managed using cooperative interference alignment with space-power adjustment and cooperative interference alignment with power adjustment [16].

8.6 Interference Estimation

Any interference in the recorded Received Signal Strength Indicator (RSSI) trace that was received from the PHY layer is what the interference prediction module is concerned with detecting. The interference estimation order to determine the current the RSSI trace's "1s" in serial and calculates interference using a threshold. First, each sequentially read RSSI value is offset by -45 before being converted to dBm. Determining the interference limit is the next step. The peak power of the receiver end without even any interference is measured to accomplish this. The signal's power level rises when it is subjected to interference because the interference is additive. The examination of the RSSI traces for different sources of interference shows that the RSSI value that occurs most frequently is the defined as the amount of the receiver side without interference. As a result, the threshold is assumed to be this value. The trace graphs also reveal that some values deviate from the threshold by less than 1 dBm.

Instead of interference from the interfering source, multipath propagation is the external component that changes these values. The next step makes use of this knowledge to convert the series into a binary trace.

Any value below the limit or any deviation from the threshold of less than 1 dBm is converted to 0. Change the remainder to 1. As a result, examples with interference are indicated by 1 and sample without interference as 0, respectively, in the binary pictures trace. The active ratio of the digital trace is computed in the module's final stage. The module predicts that the received power is subject to interference if the active ratio is larger than 0. If the active ratio is close to 0, it determines that there is no interference with the signal.

Active R(AR) = Number of samples converted as 1/Total number of sample

8.6.1 Feature Extraction

In this module, the temporal properties of the binary signal produced by the interference estimate module are sought after.

The following is a list of the features that were extracted for this project.

- Maximum time spent watching a channel
- Maximum time for channel clearing
- Channel usage percentage

A characteristic of a specific wireless transmission and reception pattern called periodicity. Two models are constructed to extract these characteristics, and the model with the lowest RAM use is chosen as the final model. The first three properties in both models are discovered using the same techniques, but the final feature, periodicity, is discovered using distinct techniques. While the lightweight technique is utilised in model 2 to calculate the frequency of the binary signal, model 1 uses the Fast Fourier Transform (FFT) to do so. Maximum channel usage duration shows the longest possible time that the signal is subject to outside interference. It can be determined by figuring out the longest time the binary signal will remain at value 1. Maximum channel clear duration reflects the longest time frame that the signal is shielded from outside interference. By calculating the length of time the binary signal can last with the value 0, it can be determined.

8.6.2 Channel Usage Ratio

The active ratio determined by the interference estimate module is the same as this. To conserve time and memory, the active ratio's value is used rather than having to calculate it from scratch. This methodology determines the periodicity by measuring the recurring switching frequency inside the binary sequence, which is a proposed simple method to estimate periodicity. The binary graph of microwave oven interference illustrate that the duty cycle consists of a regular interval of time of a busy channels and an idle channel. WiFi's binary traces analysis reveals that beacons have been sent out on a frequent basis. Even when there is data flow, it mainly revolves around the beacons. As a result, interference is present on the channel during and immediately after a beacon broadcast, but not during the interval between beacons. Bluetooth traces reveal consistent spikes that are communicated. Interference affects the channel during and after these surges. Apart from that, there is no interruption. Therefore, channel clear length and channel busy duration make up the switching frequency for each interfering technology. As a result, it is utilised to determine the periodicity or number of repeating duty cycles in a binary sequence.

Periodicity = total duration of the binary signal/channel busy time +channel ideal time

WiFi uses a 102.4 ms beacon interval by default. For WiFi, the total of the maximum channel busy and clear times should not be greater than this number. The duty cycle durations of Bluetooth and microwave ovens are consistently smaller than this figure, according to the traces. The duty cycle for WiFi surpasses this figure when there is a lot of data flow. The following criteria is utilised while computing the periodicity in order to ensure that the duty cycle consistently stays within this range. If the total of the maximum busy time and the maximum clear time is greater than 102.4 ms, the periodicity is given by:

$$Periodicity = total\ duration\ of\ the\ binary\ signal\ /102.4\ ms$$

Calculations reveal that the lightweight approach is more effective at determining periodicity. Additionally, the method's memory profile reveals that it uses less memory than FFT. Therefore, it is demonstrated that the suggested method is portable.

8.7 Machine Learning Model-Decision Tree

In order to categorise the kind of interfering signal for just an extracted feature set, the scheme suggested in this study must be used. A supervised trained model is therefore required. Having an acceptable dataset is the first criterion for creating a supervised model for machine learning. The proposed feature extraction approach is used to extract features from 70 distinct traces in order to produce the needed dataset. The process of developing an ML model is carried out after the dataset has been created. First, train data and test data are randomly selected from the dataset. The dataset is divided into 30% of test data and 70% of train data.

Data with enough observation time and updated models are needed for adequate training of the network in order to increase the generalization capability of deep learning networks [17]. The created model is validated using test data, whereas it is created and trained using training data. Two separate models are created: the logit regression model and the decision tree model. The most accurate models with the lowest memory utilization are used in the final design. The dataset's four numeric properties will be utilised as inputs for both models' four numeric parameters. The decision tree is built using the CART technique with attribute selection measures (ASM) as that of the gini index. The decision tree is more accurate and adaptable than just the logistic regression approach, according to simulation data. As a result, in the final design, the decision tree is regarded as the ML model.

8.8 Lookup Table

The look-up table selects the best countermeasure based on the type of source interference by using the result of the ML classification algorithm as its input. The table was made using observations made regarding various WSN interference countermeasures. According to these observations, each interfering technology has a unique effect on the interfered channel. A wireless network is intended to be rendered unusable by jamming and interference assaults, resulting in a denial of service [18].

For various types of interference, different countermeasures will be effective. For example, a microwave oven makes heavy use of the channel with a timely, consistent, and slow transmission. The transmission periodically goes on and off. Therefore, by delivering the packets on a timetable, or employing packet schedule transmission, the transmitting or interference periods can be avoided. So, if microwave oven interference is the entry to the look-up table, packet scheduling will be selected as the mitigation strategy. WiFi, in contrast side, uses the channel quite intensively for a very long time with a lot of data traffic. It does not have a typical on-off transmission pattern like a microwave oven. The most affordable alternative to traditional wire-lines is wireless communication, which promises to give smart grid devices a universal bi-direction information channel [19].

Therefore, switching the broadcasting channel is the best way to prevent WiFi interference. Channel switching is therefore determined to be the best preventative action when WiFi input interference is present by the look-up table. Bluetooth interferes with the WSN signal the least compared to WiFi and microwave ovens. Additionally, because Bluetooth uses an adaptive spectrum hopping mechanism to transfer data on channels, interference happens sporadically. Therefore, staying in the band and retransmitting in the event of a packet collision would be the best course of action to prevent Bluetooth interference. Using ACKs, the packet clash can be found. As a result, the look-up database will select resend with ACK as the protective measure if Bluetooth interference is found. A comprehensive introduction was made for the implementation of GNNs in wireless networks, with a focus on resource allocation and various newly emerging sectors. Several classical GNN paradigms have been utilised in wireless networks [20].

8.9 Conclusion

The deep learning methods that experts have used to reduce interference. Found crucial interference characteristics in a number of wireless communication applications. When these interference suppression strategies are compared to more conventional suppression techniques, it becomes clear that deep learning techniques have greatly improved interference suppression. Since deep learning algorithms can recognise intricate non-linear connections in the input data, they are also useful for learning complicated non-linear decision boundaries. A large dataset is necessary for deep learning approaches in order to adequately train the system and enhance test results. To improve scalability and generalizability, neural network topologies tailored for wireless networks are a potential option. It is vital to highlight the increase of interference suppression applications for deep learning. Given their distinctive qualities that can be used in interference suppression, deep learning algorithms like auto encoders & convolutional neural network are of great relevance. Graph neural networks (GNNs) are a recent attempt that can benefit from the wireless networks' graph-based architecture. Applications involving resource allocation, traffic flow prediction, and data identification have all showed promise for GNN-based algorithms.

References

1. Q. Liu, Z. Lu, and K. Yang, "Modeling and Dual Threshold Algorithm for Diffusion-Based Molecular MIMO Communications," *IEEE Trans. Nanobioscience*, vol. 20, no. 4, pp. 416–425, 2021, doi: 10.1109/TNB.2021.3077297.
2. T. Oyedare, V. K. Shah, D. J. Jakubisin, and J. H. Reed, "Interference Suppression Using Deep Learning: Current Approaches and Open Challenges," *IEEE Access*, pp. 1–1, 2022, doi: 10.1109/access.2022.3185124.
3. R. A. Seymour *et al.*, "Interference suppression techniques for OPM-based MEG: Opportunities and challenges," *Neuroimage*, vol. 247, no. November 2021, p. 118834, 2022, doi: 10.1016/j.neuroimage.2021.118834.
4. L. Liang, H. Ye, G. Yu, and G. Y. Li, "Deep-Learning-Based Wireless Resource Allocation with Application to Vehicular Networks," *Proc. IEEE*, vol. 108, no. 2, pp. 341–356, 2020, doi: 10.1109/JPROC.2019.2957798.
5. Y. Zhou, A. Samiee, T. Zhou, and B. Jalali, "Deep learning interference cancellation in wireless networks," *arXiv*, 2020.
6. S. Zhang *et al.*, "A Real-World Radio Frequency Signal Dataset Based on LTE System and Variable Channels," pp. 1–5, 2022, [Online]. Available: http://arxiv.org/abs/2205.12577
7. O. Alkadi, N. Moustafa, B. Turnbull, and K. K. R. Choo, "A Deep Blockchain Framework-Enabled Collaborative Intrusion Detection for Protecting IoT and Cloud Networks," *IEEE Internet Things J.*, vol. 8, no. 12, pp. 9463–9472, 2021, doi: 10.1109/JIOT.2020.2996590.
8. R. K. Mishra, K. Chahine, H. Kim, S. Jafar, and S. Vishwanath, "Distributed Interference Alignment for K-user Interference Channels via Deep Learning," *IEEE Int. Symp. Inf. Theory - Proc.*, vol. 2021-July, pp. 2614–2619, 2021, doi: 10.1109/ISIT45174.2021.9517735.
9. A. Suzain, R. A. Rashid, M. A. Sarijari, A. S. Abdullah, and O. A. Aziz, "Machine learning based lightweight interference mitigation scheme for wireless sensor network," *Telkomnika (Telecommunication Comput. Electron. Control.*, vol. 18, no. 4, pp. 1762–1770, 2020, doi: 10.12928/TELKOMNIKA.V18I4.14879.
10. W. Liu, "Harmonious wireless networks: Perspective of interference management," *Intell. Converg. Networks*, vol. 2, no. 3, pp. 198–204, 2021, doi: 10.23919/icn.2021.0013.
11. M. Strake, B. Defraene, K. Fluyt, W. Tirry, and T. Fingscheidt, "Speech enhancement by LSTM-based noise suppression followed by CNN-based speech restoration," *EURASIP J. Adv. Signal Process.*, vol. 2020, no. 1, pp. 1–26, 2020, doi: 10.1186/s13634-020-00707-1.
12. Z. Ma, Y. Wu, M. Xiao, G. Liu, and Z. Zhang, "Interference suppression for railway wireless communication systems: A reconfigurable intelligent surface approach," *IEEE Trans. Veh. Technol.*, vol. 70, no. 11, pp. 11593–11603, 2021, doi: 10.1109/TVT.2021.3111646.

13. W. Lee, K. Lee, H. H. Choi, and V. C. M. Leung, "Deep Learning for SWIPT: Optimization of Transmit-Harvest-Respond in Wireless-Powered Interference Channel," *IEEE Trans. Wirel. Commun.*, vol. 20, no. 8, pp. 5018–5033, 2021, doi: 10.1109/TWC.2021.3065029.
14. P. De Kerret, D. Gesbert, and M. Filippone, "Team deep neural networks for interference channels," *2018 IEEE Int. Conf. Commun. Work. ICC Work. 2018 - Proc.*, no. 1, pp. 1–6, 2018, doi: 10.1109/ICCW.2018.8403662.
15. P. Wang, Y. Cheng, B. Dong, and G. Gui, "Binary Neural Networks for Wireless Interference Identification," *IEEE Wirel. Commun. Lett.*, vol. 11, no. 1, pp. 23–27, 2022, doi: 10.1109/LWC.2021.3118903.
16. Z. Li *et al.*, "Cooperative Interference Alignment," vol. 20, no. 11, pp. 7072–7085, 2021.
17. W. Yu, H. Yu, D. Wang, J. Du, and M. Zhang, "SL-BiLSTM: A Signal-Based Bidirectional LSTM Network for Over-the-Horizon Target Localization," *Math. Probl. Eng.*, vol. 2021, 2021, doi: 10.1155/2021/9992120.
18. S. V. Manikanthan and T. Padmapriya, "Detection of jamming and interference attacks in wireless communication network using deep learning technique," 2021, doi: 10.4108/eai.7-6-2021.2308599.
19. W. Sun *et al.*, "LSTM based link quality confidence interval boundary prediction for wireless communication in smart grid," *Computing*, vol. 103, no. 2, pp. 251–269, 2021, doi: 10.1007/s00607-020-00816-7.
20. S. He *et al.*, "An Overview on the Application of Graph Neural Networks in Wireless Networks," *IEEE Open J. Commun. Soc.*, vol. 2, pp. 2547–2565, 2021, doi: 10.1109/OJCOMS.2021.3128637.

9

Implementing SDN Process-Based MRMC Wireless Networks

S. Jayasri* **and R. Parameswari**

Department of Computer Science, School of Computing Sciences, Vels Institute of Science, Technology & Advanced Studies, Pallavaram, Chennai, India

Abstract

The Multi-Radio Multi-Channel [MRMC] network's quality of services may rely on several factors, but it mostly depends on the QoS process for all the routes. Failure in ratio link leads to the reduction in performance of the QoS process as well as its throughput will also be reduced. In industries, various services are provided by accessing them with the help of the manet. Through which the employees and customers can access the sector's service anywhere, at any place, and at any time through mobile. With the node's movement in MRMC, the link-failure frequency will always be high. The nodes move in all directions with different speed limits. The rise of interest towards applying the Software-Defined Network principle of the wireless multi-hop network paves the way for bringing in flexibility and programmability in the MRMC. This is mainly lacking in the present time in the distributed wireless network such as the (Mesh, Ad-hoc, and other sensor networks). But the SDN in the wireless network promises enhanced mitigation processes like interference, mobility, and scalability of the network where the Qos service in the network gets enhanced automatically. This research tries to highlight the proposed ILP based wireless resource allocation for the p2p and the p2mp service to the end of the network with frequency needs in the SDN MRMC network hop. The recommended model's specification is based on the Wireless communication system. Also, consideration was made on switching the wireless node's resource consumption because the SDN forwarding table for the network remains restricted. The Genetic Algorithm from the proposed ILP (Integer Linear Programming) is introduced for addressing the large wireless network. The outcome from the simulation process illustrates that both the IPL and the GA from ILP formulation works effectively.

**Corresponding author*: jai.341985@gmail.com

A. Suresh, J. Ramkumar, M. Baskar and Ali Kashif Bashir (eds.) Resource Management in Advanced Wireless Networks, (183–210) © 2023 Scrivener Publishing LLC

Keywords: Access point, antenna, bandwidth, analogue phone, Wifi, Qos, genetic algorithm, LCOR, multi hop routing, multi radio multi channel

9.1 Multi-Radio Multi Channel Network in Wireless Network

In recent years, MRMC wireless networks have piqued interest.799.10a defines 14 orthogonal channels, while 802.11b/g that includes 11 channels. Amongst, three channels are orthogonal which execute under low inter-channel interference [1].

In practice, channelization has already been applied in an infrastructure mode 799.10 based wireless local area network. where adjacent APs are allocated separate channels to reduce interference and increase output. In an MR network, nodes may also have multiple network interface cards (NICs), also known as radios. Each radio is designed to switch between using several orthogonal channels, you can communicate or receive data on any of them at any moment. It communicates with many radios that perform independently over the non-interfering channels, the highest throughput and the lowest operating time in MRMC. Because of its ease of implementation and wide range of applications, a lot of published works available in multi-hop networks. The reasons behind the degraded throughput performance are the co-channel interference and the non-forecasting ability. The advancement made in the modern wireless technologies permit for different concurrent transmission with the ability of multiple radios.

With the advancement of modern wireless technologies, allowing more concurrent transmissions by using multiple radios also helps to achieve the best throughput rate. Each radio is tuned to a different frequency on a single node and the channel which is orthogonal.

Only links are placed outside of range of causing interference will communicate at the same period in wireless networks, due to co-channel interference. This has the potential to significantly increase network system ability, but it also introduces new tasks in the development of more effective RA and scheduling. Previous research in this field has mainly analyzed the link-level protocol model [1, 2]. There are also papers [3, 4] that deal with routing for such a network. The performance benefit provided by MRMC has recently been investigated, given device details like a channel or network status, radio interface, and traffic load. The efficiency is calculated in terms of the shortest total device time required to meet predetermined traffic demand. Let's us assume a routing and scheduling process of

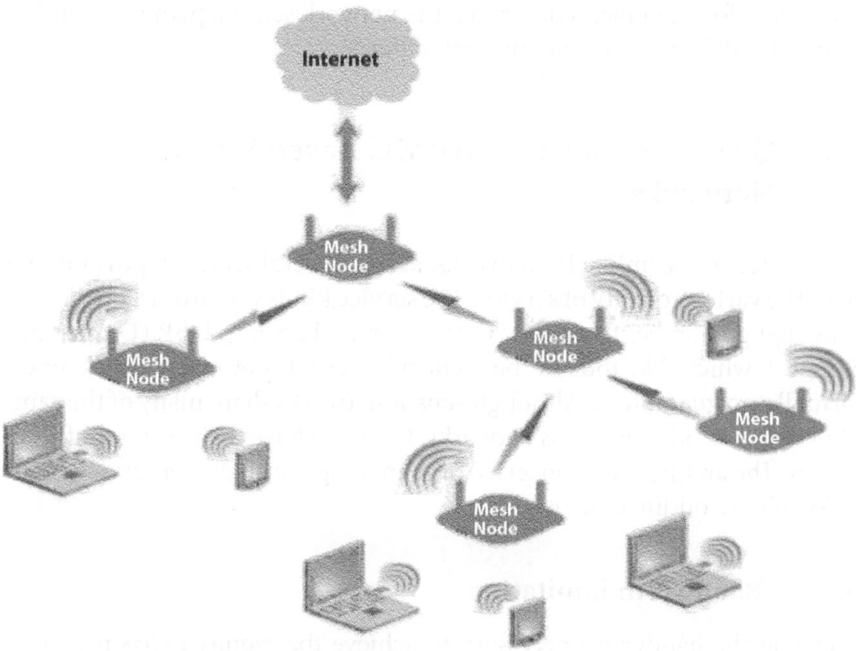

Figure 9.1 Illustrative example of MRMC.

the multi-radio multichannel that cope up with the optimization problem for route selection process.

The main goal of the MRMC is to reduce device activation time while delivering a specified traffic load over the above-mentioned MRMC and multi-hop wireless network. An omniscient and omnipotent central authority will perfectly plan packet transmission at each node. The MRMC in WN is depicted in Figure 9.1.

9.1.1 The Concept of QoS inWN

As defined in [6], A collection of service standards that the network must meet when carrying a flow is referred to as Quality-of-Service. A packet stream from a content to a destination i.e., multicast or unicast with a connected QoS [6] is referred to as a "flow." In other words, Quality of Service (QoS) is an observable level of service provided to network users, as measured by PL likelihood, usable E-E latency, frequency and other factors. Network service providers may provide such QoS in exchange for an SLA (Service Level Agreement) between service providers and network users.

Users can, for example, request that the network select a path with a minimum of 3M frequency for some traffic flows.

9.2 QoS Challenges in MRMC Based Wireless Networks

Sensor networks, unlike IP networks, accept several service types and thus have the various quality of service. The service kinds vary from Constant Bit Rate that guarantees frequency, delay jitter, and error to UBR (Unspecified Bit rate), which, like today's "best-effort" Internet protocol network, offers virtually no guarantees. Although sensor networks share many of the same QoS issues as other wireless networks, they also have their own set of challenges. The architecture concerns for managing the quality of service traffic in WSNs are outlined below.

9.2.1 Bandwidth Limitation

Securing the bandwidth necessary to achieve the required QoS is a common problem for general wireless networks. For wireless sensor networks, bandwidth constraints will become a more pressing concern. Sensor network traffic can be burst, with both non-real-time and real-time traffic. It would not be acceptable to dedicate all usable bandwidth to QoS traffic. To accommodate non-real-time traffic, a trade-off in video or image quality can be required. In addition, using multiple separate routes at the same time to break traffic and meet QoS requirements would be necessary at times.

9.2.2 Removal of Redundancy

The data generated by sensor networks is known for its high redundancy. Since simple aggregation functions will suffice for unrestrained traffic, removing terminated data messages was simple. An aggregation of data for QoS traffic, on the other hand, is even more difficult. Image and video stream comparisons are not systematically simple and can enhance a lot of energy. To render the accumulation of quality of service data systemically viable, a combination of sensor and device conditions must be needed. Since the images can be very similar, the combination of imaging data may be selectively done for traffic produced by pointing sensors in an identical direction. Another element to consider is the volume of QoS traffic at any given time. In low-traffic situations, it could be cost-effective to stop data aggregation because the overhead will become too great.

9.2.3 Energy and Delay Trade-Off

Multi Hop Routing (MHR) is almost a norm in WSN because radio transmitting capacity was relational to noisy environments or distance squared. While increasing the number of hops decreases the amount of energy used for data processing, the accumulated packet delay increases. The growth in the hops number will not only complicate the managing and analysis of latency traffic but also slow down packet delivery because packet queuing delay dominates its propagation delay. Furthermore, because of the high error rate of data routing, wireless communication may be inevitable, confusing the trade-off between packet transmission latency and energy consumption.

9.2.4 Buffer Size Limitation

The processing and storage capabilities of sensor nodes are typically restricted. In MHR, intermediate relaying nodes are used to store received packets before forwarding them to another hop. While limited buffer dimensions may theoretically suffice, in wireless sensor networks, buffering multiple packets has some advantages. To begin with, switching the radio circuitry between reception and transmission enhances an important amount of energy, so receiving many packets before forwarding them is beneficial. Furthermore, data aggregation and fusion require the use of several packets. Multi hop QoS data routing will normally necessitate larger data for buffering and long sessions, particularly when delay jitter is a concern. The buffer dimension or size restriction can enhance the delay difference that packets experience when traveling on various paths, as well as within the same path. Such a problem would hinder medium access arrangements and make meeting QoS requirements difficult.

Support for various traffic kinds: The presence of a diverse range of sensors poses a slew of technological routing problems in data. For example, few applications may need a heterogeneous set of sensors to monitor the surrounding environment's temperature, pressure, and humidity, detect motion through sound signatures and capture video or image tracking of moving objects. These specialized sensors can be organized individually or the functionality can be integrated into standard sensors for on-demand use. As previously stated, readings from these sensors may be obtained at various rates, subject to a variety of QoS constraints, and using a variety of data delivery models. As a result, data routing becomes more difficult in such a heterogeneous setting.

9.3 Opportunistic Routing

The deployment of opportunistic routing is to increase the broadcasting and the spatial diversity of the communication process using the ideology of multiple one-hop neighbors for packet forwarding. Throughput and energy efficiency both benefit from increased packet forwarding reliability. Currently available OR research focuses primarily on protocol development. The other versions such as ExOR and route forwarding depends on the costs and the candidate selection process. Depending on the cost specification, least-cost opportunistic routing (LCOR) helps to find out the optimal forwarding candidates. It make use of linear searching. Many OR versions depends on the candidate list and the relay priority that delineates the information of node's position. In the case of GeRaF, the present nodes are distinct to the zone on the basis of priority. The nodes with the highest priority moves forward to the destination. The role of data link layer elects the nodes with the hop nodes and also network layer explores the subsets of candidate nodes. To reduce packet duplication in mobile ad hoc networks, three suppression mechanisms of contention-based forwarding were studied.

9.4 The Need for Software-Defined Network in WSN for Enhancing QoS

9.4.1 QoS Management

QoS management encompasses all facets of confirming a network's appropriate service delivery. This includes device reliability and tolerance management, PL management, incident maintenance, detection, and robustness. Since the Software-defined network architecture is logically centralized, Quality service management tasks can be performed in the controller. Due to factors like energy fluctuations, topology changes, frequency availability, and node failure, the SDN-based WSN environment differs, necessitating the integration of a network reliability mechanism. [5] Models based on continuous stochastic logic techniques and continuous-time Markov chains are used to investigate the stability of the WSN framework based on SDN. The findings show that the number of sensors and controllers, as well as their respective failure rates, affects network reliability. The stability of tasks and communication networks are two types of reliability management. Methods such as having backup nodes and redundant links can be used to manage connectivity continuity [6]. The sensor functions controls the tasks by permitting the nearest nodes, which is known as self-healing mechanism. In order to enhance the task's performance, the sensor nodes imbibe with the SDN controller [7].

OpenFlow model developed for SDN based idea is described to resolve the reliability issues of wireless environment [8].

The idea is based on OpenFlow's capacity to control and track sensor traffic, which allows for secure routing and load balancing. Because the dependability of E-E communications and multi-hop in regular WSNs is still an unsolved and tough topic, the concept of sensor flow has introduced to ensure a network reliability. It is then compared with the conventional sensor flow in terms of less packet and the traffic overheads analysis [8, 9].

To control system robustness, a proposed statistical machine learning methodology must consider variables influencing and interfering with the system, as well as track interference patterns. Based on previous interference behavioral patterns, this approach will allow for the avoidance of potential interference. Tiny SDN [10, 11] has been proposed for managing latency, but further research on latency management and event detection is needed.

9.5 ILP Problem Design

This section deals with the integer linear programming (ILP) design of the VL source with the SDN relying on the multi-hop sensor field. In combination with the previous work, (1). The interventions between neighbouring links, which are limited and distributed across multiple cliques to raise the tolerance of approaching traffic, and (3) supporting the p2mp with different virtual links broadcast characteristic of wireless links, (2). The problem's variables and constraints are listed below.

9.5.1 Network Model

Here, the nodes are associated with the network interface cards under MRMC network. Characteristics of the channel assignment is static. An aggregate of || non-overlapping bands has owned a node that has q NICs, where q ||. The letter B stands for the channel potential of. the MRMC network. It was built as directed graph. It is represented as G(V, E) where V denotes the set of SDN nodes and the E denotes the set of physical links.

Each node V has a switching capacity Lv in an OF-enabled structure. The designed structue has the highest network entries. It is displayed in the flow table. *Lv* denotes the present node's size.

Table 9.1 VL source assignment models for wireless multi-hop networks.

Virtual link	P2P (any path)	P2P	P2P	P2P	P2P	P2P P2MP
VL (Quality of service)	Packet Loss Delay	Bandwidth	Bandwidth	Bandwidth	Bandwidth	Bandwidth
Multi-radio	No	No	No	Yes	Yes	Yes
Multichannel	No	No	No	Yes	Yes	Yes
Network model	Link metrics EATX and EATT	Mobility	Interference matrix	Distance depend interference design	SINR depend interference design	Conglic graph depend interference design
Method	Improved Virtual network embedding	Backtracking depend heuristic	Heuristic	GA	Heuristic and ILP	GA and ILP
Node resources	No	Storage, CPU	CPU	None	CPU	FT and group entries
Spliting the paths	No	Not supported	No	No	Yes	Yes for ILP and no for GA

An OF table is also taken into account. The simulated packets from virtual connection or unique connection are used for path splitting process which is displayed in GT table.

Similarly, Mv and Mv′ denote the determined and existing sizes of node v's group table, respectively. We're going to say that we've already had a successful channel allocation . ζ. ζ assigns a collection of $v \in V$ a set of ζ (v) of $|\Lambda|$ separate channels to each node $\zeta (v) \subseteq \Lambda$.

If two NICs are on the same channel and within the transmitting range of each other, they can communicate with each other. The wireless communication $e = ((v,u),\lambda)$, $u,v \in V$ and $\lambda \in |\Lambda|$ belongs to the substrate network if channel $\lambda \in \zeta (v) \cap \zeta (u)$ and nodes v and u are within routing path of each other on this latter channel. N denotes the set of v's neighbors (v).

9.5.2 Interference Model

The definition of conflict graphs [11] underpins our intervention model. A channel is linked to a conflict graph. When both links (vertices) are active at the same time, it defines the existence of interferences. We derive maximal cliques [13] from the conflict graph, adhering to the logic of earlier works [12]. A clique is a conflict graph subgraph in which all nodes affect one another. Most cliques are used in a variety of ways to originate various constraint types on the communication speeds of the wireless links that make up the clique [14] and [15].

There are several methods for creating conflict graphs, each with varying degrees of precision in capturing interference. We do not advocate for any approach in this paper, and we believe that we have the maximum struggle graphs as data. The essential graphs that fulfil the requirements of the network control process may be accessible by the controller due to the unified nature of the software-defined network and the deployment of efficient service at the controller. The group of maximal cliques is referred to as C in the following. The pool of links that uses clique c is referred to as Ec. Under the substrate network, it is represented as Sc.

9.5.3 VL Request Design

A VLR request is made up of several VL called $|K|$. Each virtual connection $k \in K$ has the following characteristics:

- a origin node $sk \in V$ with the set of destination nodes. The destination nodes are represented as $Tk \in V \backslash sk$ (the VL follows direct communication approach when $Tk = 1$, otherwise it is p2mp); • a network throughput K® = [K1,K2,...,] denotes the sequence of virtual connection requests.

9.6 Problem Constraints

The bandwidth allocation constraints are defined in equations 9.1 to 9.9 below. Inequalities 1 and 2 describe the restrictions on switching resource allocation. It includes group table and the network flow details that increases the information of group table with the available sensor nodes. The preserveration of group table is to assess and monitor the entries of sensor flows that is being allocated during virtual link connection not increase the total of group table and flow available on the sensor servers.

The linearization of the *Min* and *Max* and operator added to the variables *lk*(*v*) to obtain l_{max} and l_{min} is shown in Equation 9.3.

$$\forall v \in V: \sum_{k \in K} l_k(v) \leq L_v - L'_v \tag{9.1}$$

$$\forall v \in V: \sum_{k \in K} m_k(v) \leq M_v - M'_v \tag{9.2}$$

$$\forall v \in V: l_{min} \leq L_v - L'_v + \sum_{k \in K} l_k(v) \leq l_{max} \tag{9.3}$$

Whatever the destination, Constraint 4 represents the frequency linearization f t (e) allotted to VL k at connection e = ((uv),λ). 5th equation guarantees that the overall frequency allocated to the clique's substrate wireless nodes does not surpass the clique's remaining bandwidth. Each maximal clique with its connected channel was denoted as (c,λ) ∈ C, which was collected of Ec, and the residual capacity was denoted aξ(c) = Bλ – ξ ′(c), with ξ ′(c) indicating frequency allocations.

$$\forall k \in K, \forall e = ((v,u),\lambda) \in E, \forall t \in T_k : f_k^t(e) \leq f_k(e) \tag{9.4}$$

$$\forall (c,\lambda) \in C: \textstyle\sum_{k \in K} \sum_{v \in S(c)} f_k(v,\lambda) \leq B_\lambda - \xi'(c)(c,\lambda) \in C: \xi_{min} \leq \\ B_\lambda - \xi'(c) + \textstyle\sum_{k \in K} \sum_{v \in S(c)} f_k(v,\lambda) \leq \xi_{max} \tag{9.5}$$

$$\forall k \in K, \forall t \in T_{k+\forall v \in V:} \sum_{u \in N(v) \lambda \in \zeta(u) \cap \zeta(v)} \left(f_k^t(e_1) - f_k^t(e_2) \right) = \begin{cases} b_k & \text{if } v = s_k \\ -b_k & \text{if } v = t\ e_1((v,u),\lambda) \\ 0 & \text{else} \end{cases} \tag{9.6}$$

$$e_z((u,v),\lambda) \tag{9.7}$$

fk((v,u),) and дk((v,u),) are channeling constraints between integer and binary variables. It also limits the bandwidth assigned to the VL k over aexterior association of the demanded throughput bk. Equation 9.9 dictates the binary variables of channel tunneling process, as well as a bandwidth constraint for packets destined for given virtual links end-point. The right-hand inequality ensures that the VL's bandwidth requirement is further analyzed. The variables at left side determines the path-splitting process during the assignment of bandwidth. The splitting of paths is processed during the allocation of bandwidth that meets the threshold constraints, bk^{min}.

In practice, *bkmin* is a ratio of *bk*, $bk^{min} = PS_{ratio} * bk$ with $PS_{ratio} \in [0,1]$ (then, $PS_{ratio} \leq 0.5$ when the path-splitting was acceptable, and $PS_{ratio} = 1.0$ when it is terminated).

$$\forall k \in K, \forall v \in V, \forall u \in N(v), \forall \lambda \in \zeta(v) \cap \zeta(u): g_k((v,u),\lambda) \leq f_k((v,u),\lambda) \leq b_k * g_k((v,u),\lambda) \tag{9.8}$$

$$\forall k \in K, \forall t \in T_k, \forall v \in V, \forall u \in N(v), \forall \lambda \in \zeta(v) \cap \zeta(u): b_k^{min} * g_k^t((v,u),\lambda) \leq f_k^t((v,u),\lambda) \leq b_k * g_k^t((v,u),\lambda) \tag{9.9}$$

9.6.1 Genetic Algorithm

Here ILP-based algorithm was used to find exact solutions to the problems under consideration. The computational complexity grows exponentially with the no. of channels, connections, parameters, and nodes which may make large networks practically unfeasible. Then present a GA-based approach to address this issue in this section. Algorithm 9.1 describes the overall workflow of the genetic algorithm scheme.

Algorithm 9.1: GA-based resource allocation.

Input : $G(V,E); K; W = [we_1, we_2,..., w_{e_{|E|}}];$
$\alpha_1; \alpha_2; \alpha_3; \beta_1; \beta_2; \beta_3; N_p; N_g; cxPB; mutPB;$

Output : $\chi = [\tau_1, \tau_2,...\tau_{|K|}]$ (i.e. the best individual)

```
begin
    P0 ← InitialPop(G, K, Np ,W)
    P ← P0
    for ( jg = 0; jg < Ng; jg ++) {
        for ( jp = 0; jp < Np; jp ++) {
            (χa, χb) ← TournamentSelection(P)
            χc ← Crossover (G,K, (χa, χb), cxPB, W)
            P ← P ∪ Mutation(χc, mutPB)
        P ←
          PopulationSelection(P, Np, α1; α2; α3; β1; β2; β3)
    x ← SelectionBestIndividual(P)
```

9.6.2 Balanced Source Distribution with DL Cost

e = α1 + α2 + β1 E(c) for the standard connection cost of e = ((v,u)) (λ). If this static method of crucial connection cost is applied to all requests in K® = [K1,K2,...], that dictates the cliques on low links based on the estimated loads. With the use of clique balancing process, the evaluation of fitness function has analyzed on generating the candidate solutions. To overcome this, Algorithm 9.2 shows a dynamic version of connection cost. If a clique c's clique utilization is computed with top N loads, then the cost of c's will be multipled with 2.1 as per algorithm. In case clique c is used for embedding process of K (i.e. χK ∩ c ,∅„ as seen in Line-11 of Algorithm 9.2), then links in c should be multiplied by 1.5 before the Knext embedding solution is calculated. As a result of their high connection costs, the most frequently used cliques would be disfavored in the implanting of future requests. Note that link cost increases will be compounded over time.

If a clique remains among the topmost loaded from Ki to Ki+Δ, then cost of the link is increased to +1 times before the computation of top loaded list. The cost of the clique's links will be performed until N-top is found.

9.7 Virtual Network Embedding

In the last few years, virtual network embedding has gotten a lot of press. Most embedding strategies have look for different substrate networks. These processes combines with different set of elements that provides route path set of choices. It is also believed that the successful P2MP links in wireless multihop network. Addition to that, the SDN based multi-hop network is new concept that has many switching resources. The development of multi-cast environment under different interference mode is been used to reduce the use of resource consumption and also to enhance the reuse of spatial information.

9.7.1 Specification of Virtual Network Embedding

In the embedding process, the functioning of wired network infrastructures is different. The deployment of wireless medium has renowned the radio resource usage with some interference limits. The node connected with virtual links are repliced until the nodes are converged and communicated.

The single transmission of red channel under different interfaces are equipped with different nodes. In order to reduce the radio resources, the communication criteria looks for different aspects of wireless links. To reduce radio resources, an effective embedding method should use the broadcast aspect of wireless communications whenever possible. On the other hand, if we examine a wireless network with many radio nodes and various operational channels, there are several options for supporting the P2M VL. It s a single-hop data that passes by red or green channel. Whereas, bi-hop data path make use of yellow and red channels. These choices have varying effects on the neighboring network. Drastically, the transmission of red channel that rely on the broadcast of green channel. To encourage spatial reuse of radio channels, the green channels are used to broadcast. To make the best use of radio resources, an effective embedding technique should minimise interference on nearby lines whenever possible.

9.7.2 System Model

The wireless embedding system consists of a substrate network and multiple virtual networks. Usually, a substrate network is defined as a weighted undirected graph. In wireless Virtual Network Embedding process rather than specific substrate node and link resources, such as computation capability, storage memory and link bandwidth.

9.7.3 Open Flow Enabled Network

We'll look at a network with SDN/OpenFlow capabilities. The node with OpenFlow follows different group table associated with the different entry. Each flow is interconnected with the different flow table entry. It also merges with the match rules that recognizes the packet forwarding actions. In some cases, the counters make use of different flow statistics. These are further combined with the different flow table. The GT entry composes of node's information obtained from the fault-tolerant networks and the load-balancing algorithms. It is composed of different buckets where each bucket has been serving backup model to one another.

9.7.4 Network Model

One or more Network Interface Cards are installed in different nodes with wireless MRMCs. The efficient use of radio resources are also tune up with the channel that coordinates NICs. We'll presume that the channel assignment is predetermined and unchanging. Let B be the channel capacity.

In the system, there are a total of || non-overlapping frequency channels, and each node has q NICs, where q ||. The MRMC network. It was built as directed graph. It is represented as G(V, E) where V denotes the set of SDN nodes and the E denotes the set of physical links. Let v be the present node's size with flow table information. It is explored in the openflow group table. The path splitting process has renowned the multiple interfaces with different links.

Similarly, Mv and M'v denote the maximum and current sizes of node v's group table, respectively. We're going to presume that we've already gotten a good channel assignment. Assigns a collection of (v) of || distinct channels to each node v V: (v). If two NICs are on the same channel and within transmission range of one other, they can communicate with one another. In other words, if channel (u)(v) and nodes u and v are within transmission range of each other, the wireless link e = ((u, v),), u, v V, and || belongs to the substrate network. N(v) refers to the collection of v's neighbours (across all channels).

9.7.5 Interference Model

The concept of conflict graphs underpins our interference model. A conflict graph expresses the association with different pair-wise interferences under different frequency ranges. The association of multiple links forbids the MAC protocols. Furthermore, the SDN based wireless network include six nodes and the conflict graphs are shown for different illustration purposes. In order to develop a conflict graphs, various techniques are employed to efficiently use wireless links, in order to eliminate the interferences issues. On the other hand, two links are shared on nodes with the similar frequency. The broadcast on such nodes are not possible. The wireless modes are from the interference range that can hold different possibilities.

9.8 Algorithm on Interference Modeling and Channel Selection Process

To make it easier to choose the proper set of resources. Here, each path is located and added progressively. As a result, we offer a routing and channel selection method that helps to design routing algorithm for interference-aware. It helps to identify the path.

9.8.1 Interference Aware Routing Algorithm

The deployment of routing aware algorithm from all unselected sources with the highest MIPC. To achieve this purpose, we present the distributed routing process that imbibes the interference-eliminationby following MIPC metric, as illustrated in Algorithm 9.1, which is based on AODV routing. Other sources are unselected, except the selected source set with the R. By broadcasting a route request (RREQ) message, the highest MIPC make use of different path's hop. Whenever an intermediate nodes are performing, the broadcast appears by receiving the RREQs, IDs, selected wireless links and the channel process. The path capacity of the nodes are also assured between the source and the destination. Relied on the destination's position, the received request from source are analyzed. When a destination receives numerous RREQs from a source, it chooses the one with the highest MIPC and uses that path as the source's found path. The destination then chooses the path with the highest MIPC as the option from the set of sources not used.

Step 1: The non-selected source s creates RREQ messages with its ID, capacity Cp(s,s), and selected source set R. (line 1). Cp(s,s) is abbreviated to Cs. The capacity of the path source to source is given as Cs, which has a value of $+\infty$.

Step 2: Based on received RREQ message from source node, it eliminates the data received from same node and discard for further communication.

Step 3: Likewise, an intermediate node x receives RREQ messages from other nodes, the channel link e(y, x) are used for communication. The available capacity Ce(y,x) for link e(y, x) is then calculated using, and the path capacity from the source s to itself is calculated using Cx = min Ce(y,x), Cy. Here, Cy is the capacity of the path between s to y nodes. In some cases, the messages are rebroadcasted with its Ids, channel Chk, and Cx to the RREQ. The RREQ is discarded in the case of circular path.

Step 4: When the destination node d gets an RREQ from node y, Algorithm 9.2 determines the channel Chk of the link between the two nodes. It then uses (4) to calculate the available capacity Ce(y,d) of link e(y, d) and Cd = min Ce(y,d) to calculate the path capacity from source s to destination d, Cy. Finally, it uses MIPC (p) = Cd jp IFIj to calculate the MIPC metric of the path that corresponds to the RREQ message. The destination node will receive several RREQ messages, each corresponding to a candidate path, because there are many candidate paths from multiple origins to the destination. From all of the possible paths, the destination node should choose the path with the highest MIPC. As a result, the first time the destination node receives an RREQ message, it sets a timer. When the timer expires, the destination chooses the path psi with the highest MIPC out of all the RREQ messages received from source si, then chooses the path with the highest MIPC out of all psi. After that, it sends an RREP message to all of the source nodes, with the ID and channel of all of the nodes along the path.

Step 5: The received reply from intermediate node x from node y checks for channel allocation that links two paths from upstream and downstream nodes. If yes, the allocated channels for two links perform based on the node ID. These are then forwarded to the RREP message. If this is not the case, the RREP is discarded.

Step 6: The source node s checks to see if it is included in the RREP message after receiving it. If so, it selects itself and assigns a channel to its radio interface based on the RREP channel list. If this is not the case, source node s keeps track of the source that was chosen.

9.8.2 Channel Assignment Algorithm

Using RREQ and RREP messages, we discovered the path with the highest MIPC. The node x must selects the channel link with upstream node y by receiving the RREQ message from node y. The interference range of link with reduced wireless interference of different throughput are presented. The nodes with similar transmission range are operated on the communication.

The working channel set, represented by CH, is the set of channels that a node's radios are working on (x). The accessible channel set of node x is defined as the channels that the node can work with, represented by ACH (x). The allocated nodes with the available channel set make use of orthogonal that can be explored.

The channel for the link e(y, x) should be at the intersection of the x and y accessible channel sets; that is, ACH(x) ∩ ACH(y). This channel set is named the feasible channel set and is denoted by FCH(y, x) = ACH(x) ∩ ACH(y). As a result, the goal of our channel assignment problem is to find the channel in FCH that has the least amount of interference (y, x). The channel assignment algorithm is presented in Algorithm. Lines 1 and 2 show how the link's two end nodes calculate the available channel. The feasible channel set in line 3 is calculated by node x. When multiple links in a local region use the same channel, intense competition occurs, resulting in poor capacity. As a result, from all the possible channels, we chose the one with the fewest working links as the working channel to minimise interference.

9.8.3 The MCM Algorithm

The information about overlapping and orthogonal channels are considered in MCM algorithm. Interference factors are used to reduce interference between one-hop nearby nodes. Thus, the working of MCM-CA technique is given. The source code is responsible for the connection links that differentiates the downlink interfaces. The source node's child node listen to the channel information received from the source node. By the BFS process, the algorithm is performed for the source nodes. The downlink interface is also interfaced with single-hop neighbors.

Let N(v) be the set of node v's one-hop neighbours; cv denotes the channel assigned to node v; and (cv,cw) denotes the interference factor between two channels cv and cw. The deployment of MCM with different forwarding nodes are used by formulating the minimized function:

$$\forall w \in N(v)\ \delta 2\ (cv,cw)$$

If there are numerous channels that satisfy via optimization function that chooses the random of all different options. The process of BFS traversal, the channel allocation determines until different forwarding nodes. The random position of MCM will determined from breadth first search traversal. Meanwhile, CA operation will be covering all forwarding nodes.

9.9 Performance Evaluation

The concept of review analysis is to describe the models approximately by focussing on three links. It also compares and contrasts the two proposed approaches, as well as the trade-offs higher by the computation time Vs latter accuracy.

9.9.1 Network Model

Only one network instance was measured in the results. It is made up of 30 nodes linked by 70 connections. The nodes have three radio interferences executes on six different channels. Each channel has a capacity of 190 UB. The network topology is depicted on the left side of Figure 9.2, with each connection color representing a band. It focuses on nine cliques. The group and flow table maximum sizes are arranged to 100 and 1000, respectively, unless otherwise specified.

9.9.2 Load Design Algorithm

Virtual connection requests are made up of a collection of 4 to 6 p2p and p2mp VLs that are selected at random. A destination number was selected randomly between 3 and 9 for each p2mp connection. Each virtual link's bandwidth requirement was chosen at random from 2 to 4 UB. The selection of the source and destination is done at random. The deliveries of requests make use of poisson method. The incoming rate r with intervals, 0.02, 0.04, 0.06 and 0.08 per 120 Uts.

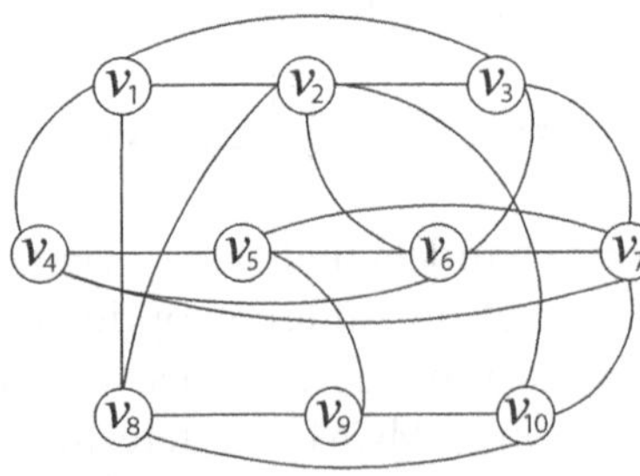

Figure 9.2 Example of conflict graphs.

Algorithm 9.2: Balanced source distribution with FL cost.

Input : $G(V, E)$; $\vec{\mathbf{K}}$; C; N_{top}; $\alpha_1, \alpha_2, \alpha_3, \beta_1, \beta_2, \beta_3, N_p, N_g, cxPB, mutPB$

begin
- **foreach** $e \in E$ **do**
  - $W[e] \longleftarrow \alpha_1 + \alpha_2 + \beta_1, |E(c)|$
- **foreach** $K \in \vec{K}$ **do**
  - $X^K \longleftarrow$ geneticAlgorithm $(G, K, W, \alpha_1, \alpha_1, \alpha_2, \alpha_3, \beta_1, \beta_2, \beta_3, N_p, N_g, cxPB, mutPB)$
  - $C_{most} \longleftarrow$ mostUsedCliques (C, N_{top})
  - $C_{least} \longleftarrow$ leastUsedCliques (C, N_{top})
  - $C_{normal} \longleftarrow C - C_{most} - C_{least}$
  - **foreach** $c \in C_{most}$ **do**
    - **if** $c \cap X^K \neq \emptyset$ **then**
      - **foreach** $e \in c$ **do**
        - $W[e] \longleftarrow W[e] \times 1.5$
    - **else**
      - **foreach** $e \in c$ **do**
        - $W[e] \longleftarrow W[e] \times 1.1$
  - **foreach** $c \in C_{least}$ **do**
    - **if** $c \cap X^K = \emptyset$ **then**
      - **foreach** $e \in c$ **do**
        - $W[e] \longleftarrow W[e] \div 1.5$
    - **else**
      - **foreach** $e \in c$ **do**
        - $W[e] \longleftarrow W[e] \div 1.1$
  - **foreach** $c \in C_{normal}$ **do**
    - **foreach** $e \in c$ **do**
      - $W[e] \longleftarrow \alpha_1 + \alpha_2 + \beta_1, |E(c)|$

9.9.3 Simulation Settings

The CPLEX 12.63 solver was used to implement the IL design in Python. The tests were carried out on an Ubuntu 14.04 virtual machine with a 30 central processing unit and 18 gigabytes of random access memory. A difference of less than 3% from the ideal solution was measured appropriately. The routing path was inactivated for integer linear programming unless it is explicitly enabled. The GA implementation is written in Python (pypy 1) and uses deep [16, 17]. Unless otherwise mentioned, the population size and number of generations are set to 18 and 18 respectively. The probability of cross-over is 0.9, and the probability of mutation is 0.05. The value of Ntop is set to 2. 10000 UT is set as the simulation horizon. Throughout all evaluation experiments, . $\alpha 1$, $\alpha 2$, and $\alpha 3$ are set to 1, 1, and 5 respectively.

9.9.4 Performance Metrics

For performance analysis, the corresponding process metrics are calculated.

1. Ac: the number of active virtual connection response to the received requests under stipulated period.
2. Clique utilization is defined as the bandwidth allocated to the links that make up a clique divided by the channel size.
3. Node usage with flow table: It collects the source table information based on received requests.
   The size of the collected source table will determine the utilization of the received requests.
4. In terms of group table consumption, switch source utilisation is high. (du, in percent): node group table utilization separated by the size of the group table.
5. Average calculation time (in seconds): the time it takes to complete one request.

9.10 Performance Results

9.10.1 Handling with WL Intervention

The aim is to see how effective our methods are at minimizing and preventing interference in wireless links. β2and β3 are set to 0 in the first position to do this. The effect of setting β1 to 1 versus 0 is compared. Figure 9.3 depicts the use of two types of cliques: The small cliques with interfering links and the large cliques with interfering links are also studied.

When interference reduction is allowed, large cliques utilize less bandwidth and small cliques utilize more.

9.10.2 Evaluating the Multicast Gain

The aim was to measure the source savings achieved by the proposed model that can use of multicast advantage. This embeds the multipoint links of virtual connection. In the first position, β2 and β3 are set to 0.1 is set to 1 since we've established that interference should be taken into account while modeling. Figure 9.4 depicts the 9 cliques' use of cliques. Disabling

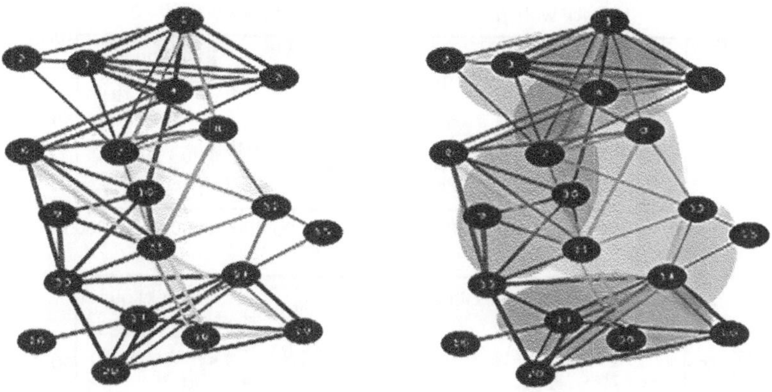

Figure 9.3 Network model.

the multicast advantage results in increased bandwidth usage, which overloads all cliques and leads to a large increase in embedding failures.

9.10.3 Clique Utilization Balancing

Clique load balancing is allowed by setting β2 to 5. Figure 9.5 indicates that, as opposed to Figures 9.3 and 9.4, where $\beta_2 = 0$. The use of clique is desparated with the GA and ILP. This can increase the 99.5% ILP and 98.5% GA of acceptance rate.

9.10.4 Analysis of Switch Resource Consumption

The usage of flow table and its analysis are described as follows.

To demonstrate how FT source was balanced and consumed, it set the initial flow table size to 80. It is then compared the results of β3 = 0 versus β3 = 0 using ILP, as seen in Figure 9.6. Then see a higher acceptance rate, in addition to a far better resource balance in the flow table use (99.5 percent Vs. 94.5 percent). As a result, resource balancing for flow tables should be allowed. Figure 9.7 depicts the result of GA switch resource balancing. In terms of switch resource balancing, the genetic algorithm was less efficient than ILP. Figure 9.8 shows the community table consumption, which was calculated using ILP and GA. Despite the efficient mapping of p2mp virtual links, we observe that group table consumption remains very

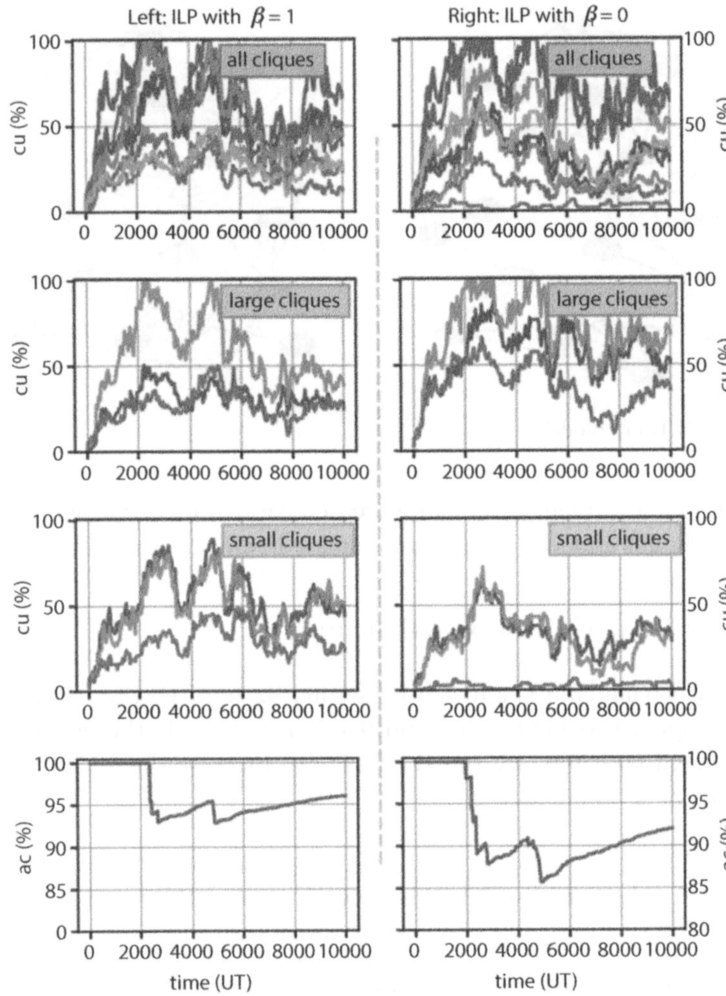

Figure 9.4 Acceptance rate for clique utilization with $\beta_1 = 2$, v.s $\beta_1 = 3$.

low with GA and ILP due to the multicast benefit. As predicted, GT entries are plentiful in comparison to embedding requirements for the simulation model under consideration. As a result, it does not play a significant part in the data path collection. As a result, there is no need to balance its use in the formula.

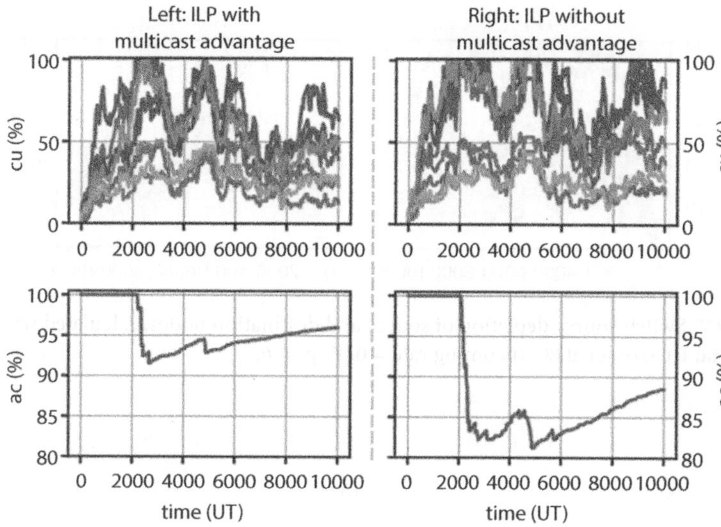

Figure 9.5 Clique consumption and acceptance rate (Arrival rate = 0.03. $\beta_2 = 1$, $\beta_3 = 10$).

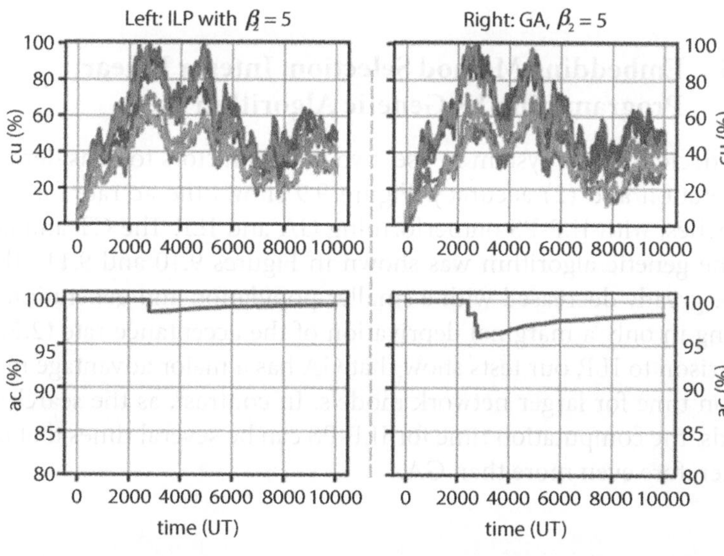

Figure 9.6 Clique consumption computing GA and ILP. Incoming rate = 0.05. $\beta_3 = 10$.

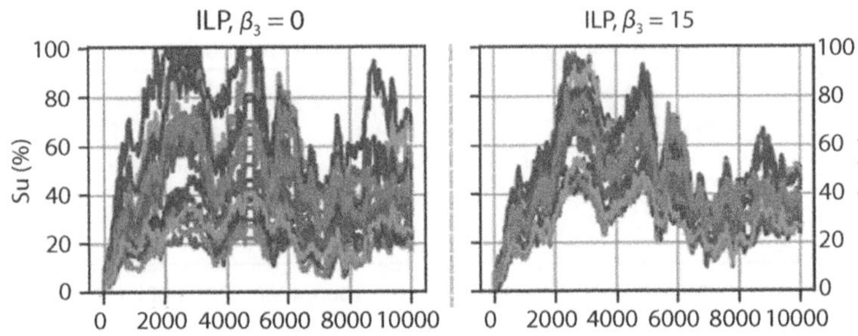

Figure 9.7 Switch source depletion of source and destination nodes, calculated with ILP, with initial FT size set at 80. Incoming rate = 0.05. $\beta_2 = 6$.

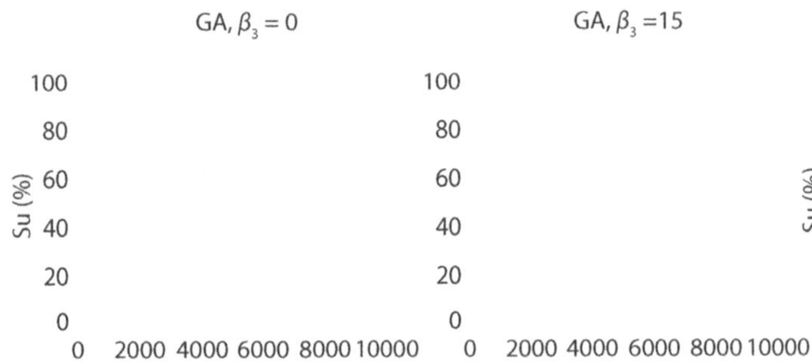

Figure 9.8 Switch source depletion of all nodes, calculated with GA, with initial FT size set at 80. Incoming rate = 0.06. $\beta_2 = 6$.

9.10.5 Embedding Method Selection: Integer Linear Programming Vs Genetic Algorithm

When deciding on a system to use, two critical factors to consider are (1) Computation and (2) accuracy. Figure 9.9 shows the ac rate for various approaches, with ILP-PS outperforming GA and ILP. The CT and ac rate with the genetic algorithm was shown in Figures 9.10 and 9.11. The CT may be greatly decreased with a smaller population and generation sizes, resulting in only a marginal deprivation of the acceptance rate (2.5%). In comparison to ILP, our tests show that GA has a major advantage in computation time for larger network models. In contrast, as the search space expands, the computation time for ILP-PS can be several times that of ILP, and therefore even more than GA.

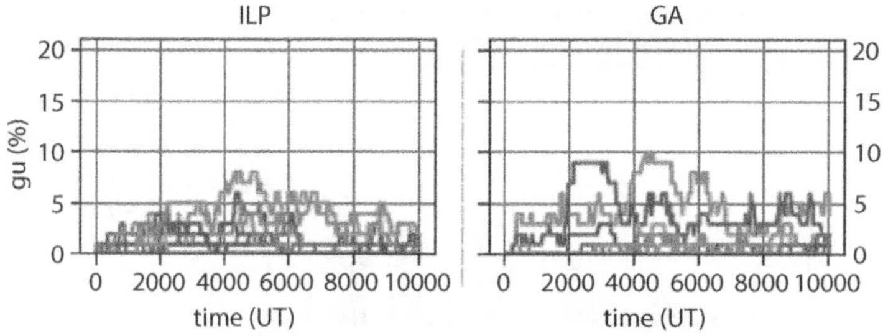

Figure 9.9 GT analysis of GA & ILP (Incoming rate = 0.32. $\beta_2 = 6, \beta_3 = 10$).

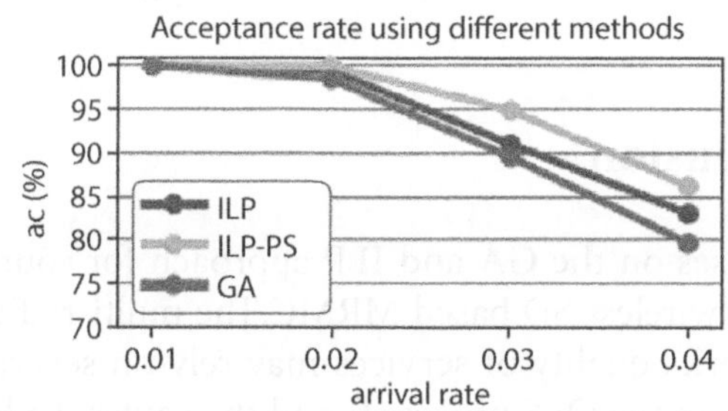

Figure 9.10 Acceptance rate with ILP-PS, ILP and GA ($\beta_2= 6, \beta_3 = 10$).

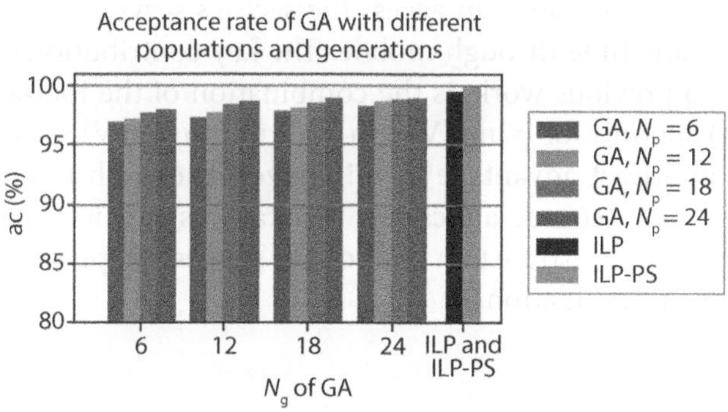

Figure 9.11 Acceptance rate of genetic algorithm by different population size and generation compared with ILP-PS and ILP.

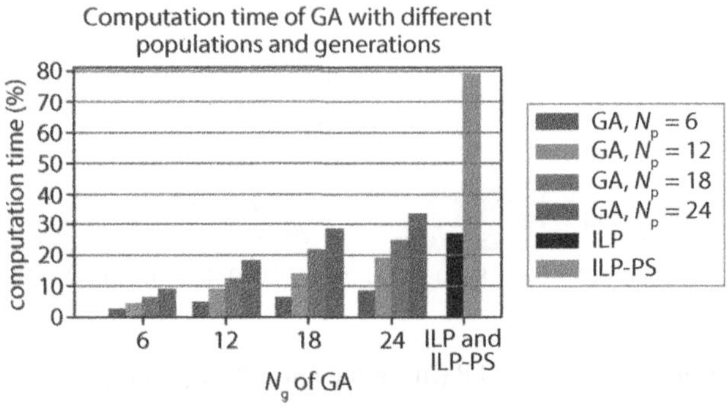

Figure 9.12 ACT of genetic algorithm by different size and generation compared with ILP-PS and ILP.

9.11 Conclusion

This paper focuses on the GA and ILP approach for source allocation of different VLs in wireless SD based MRMC.The multi-radio multi-channel [MRMC] network's quality of services may rely on several factors, but it mostly depends on the QoS process for all the routes. Failure in ratio link leads to the reduction in performance of the QoS process as well as its throughput will also be reduced. In industries, various services are provided by accessing them with the help of the manet. Through which the employees and customers can access the sector's service anywhere, at any place, and at any time through mobile. The key contribution of our ideas, in contrast to previous work, is the combination of the following characteristics: (1) support for p2mp VLs in addition to p2p VLs, as well as how to use the multicast advantage to minimize bandwidth usage in a wireless context; (2) resource allocation that takes switching resources into account. Then demonstrate that both of our suggested approaches perform well through our evaluations.

References

1. P. Bahl, R. Chandra and J. Dunagan, "SSCH: Slotted seeded channel hopping for capacity improvement in IEEE 802.11 ad hoc wireless networks," Proc. IEEE MobiCom'04, 2004.
2. N. Jain and S.Das, "A multi channel CSMA MAC protocol with receiver based channel selection for multihop wireless networks," Proc. IC3N'01, 2001.

3. R. Draves, J. Padhye and B. Zill, "Routing in multi-radio, multi-hop wireless mesh networks," Proc. ACM MobiCom'04, 2004.
4. A.Raniwala, K.Gopalan and T.Chiueh, "Centralized channel assignment and routing algorithms for multi- channel wireless mesh networks," ACM MC2R, vol. 8, no. 2, April 2004.
5. Lu, Y.; Huang, X.; Huang, B.; Xu, W.; Zhang, Q.; Xu, R.; Liu, D. A Study on the Reliability of Software Defined Wireless Sensor Network. In Proceedings of the 2015 IEEE International Conference on Smart City/SocialCom/SustainCom (SmartCity), Chengdu, China, 19–21 December 2015; pp.129–134.
6. Zhou, J.; Jiang, H.; Wu, J.; Wu, L.; Zhu, C.; Li, WSDN-based Application Framework for Wireless Sensor and Actor Networks. IEEE Access **2016**, 4,1583–1594.
7. Miyazaki, T.; Shitara, D.; Kawano, R.; Endo, Y.; Tanno, Y.; Igari, H. Robust Wireless Sensor Network Featuring Automatic Function Alternation. In Proceedings of the 2011 20th International Conference on Computer Communications and Networks (ICCCN), Maui, HI, USA, 31 July–4 August 2011; pp.1–6.
8. Mahmud, A.; Rahmani, R. Exploitation of OpenFlow in wireless sensor networks. In Proceedings of the 2011 International Conference on Computer Science and Network Technology (ICCSNT), Harbin, China, 24–26 December 2011; Volume 1, pp.594–600.
9. DeOliveira, B.T.; Margi, C.B.; Gabriel, L.B.TinySDN: Enabling multiple controllers for software-defined wireless sensor networks. IEEE Lat. Am. Trans. **2015**, 13,3690–3696.
10. Orfanidis, C. Ph.D. Forum Abstract: Increasing Robustness in WSN Using Software Defined Network Architecture. In Proceedings of the 2016 15th ACM/IEEE International Conference on Information Processing in Sensor Networks (IPSN), Vienna, Austria, 11–14 April 2016; pp.1–2.
11. DeOliveira, B.T.; Margi, C.B.; Gabriel, L.B. TinySDN: Enabling multiple controllers for software-defined wireless sensor networks. In Proceedings of the 2014 IEEE Latin-America Conference on Communications (LATINCOM), Cartagena de Indias, Colombia, 5–7 November 2014; pp.1–6.
12. M. Capelle, S. Abdellatif, M. J. Huguet and P. Berthou, "Online virtual links resource allocation in Software-Defined Networks," IFIP Networking Conference (IFIP Networking), Toulouse, 2015, pp.1-9.
13. Chen, Lunde, Slim Abdellatif, ArmelFrancklin Simo, and Thierry Gayraud. "Virtual link embedding in software-defined multi-radio multi-channel multi-hop wireless networks." In Proceedings of the 21stACM
14. International Conference on Modeling, Analysis and Simulation of Wireless and Mobile Systems, pp.163- 172. 2018.
15. Zhang, Jihui, Haitao Wu, Qian Zhang, and Bo Li. "Joint routing and scheduling in multi-radio multi-channel multi-hop wireless networks." In 2nd

International Conference on Broadband Networks, 2005., pp. 631-640. IEEE, 2005.

16. Liu, Yunxin, YongqiangXiong, Yang Yang, Pengzhi Xu, and Qian Zhang. "An experimental study on multi-channel multi-radio multi-hop wireless networks." In GLOBECOM'05. IEEE Global Telecommunications Conference, 2005., vol. 1, pp. 5-pp. IEEE,2005.
17. Ndiaye, Musa, Gerhard P. Hancke, and Adnan M. Abu-Mahfouz. "Software defined networking for improved wireless sensor network management: A survey." Sensors 17, no. 5 (2017):1031.

10

Advanced Wireless Mobile Network on Financial Literacy

R. Muthu Lakshmi[1]* and A. Maria Bharathi[2]

[1]*Sri Sarada College for Women (Autonomous), Tirunelveli, Tamil Nadu, India*
[2]*Department of Commerce, Sri Sankara Bhagavathi Arts and Science College, Kommadikottai, Tuticorin, Tamil Nadu, India*

Abstract

Advanced Wireless network plays an important role to develop financial inclusion and financial literacy. Financial literacy can be improved through wireless network among the people. Digitalisation becomes life blood of our movement. Providing opportunities for learning in a digital arrangement can therefore engage individuals and entrepreneurs in modes of learning that mirror their personal activities. Technical financial literacy is likely to become aprogressively important aspect of education. The growth of the digital economy means that persons will become more accountable for their own monetary planning. In the modern world, Peoples will need to manage their funds individually with available financial knowledge concerned with savings and investment plans. Consumers and all the people will need to have a greater level of financial cleverness to make effective use of monetary technology products and services and avoid fraud and expensive faults. Digital monetary education programs will improve digital financial literacy among the people, students and Investors.

Keywords: Financial inclusion, financial literacy and wireless network

**Corresponding author*: muthulakshmisarada@gmail.com

A. Suresh, J. Ramkumar, M. Baskar and Ali Kashif Bashir (eds.) Resource Management in Advanced Wireless Networks, (211–226) © 2023 Scrivener Publishing LLC

10.1 Introduction

During this COVID - 19 pandemic, digitalization plays an important role in all areas. Because it ensure that people stay at home with satisfaction and reduce the unnecessary movement. Government also wants to provide their services to the people electronically by improved online infrastructure.

Activities like capacity to realise and efficiently apply various monetary skills, including ownmonetary management, budgeting, and investing etc. comes under financial literacy. Through financial literacy individuals can become independent so that they can attain financial steadiness.

Digital Monetary Services (DFS) ensure that enhancement of financial inclusion and financial literacy. With the help of wireless mobile network, banks and non-banks can provide digital financial services to the people who are live in rural areas.

10.2 Statement of the Problem

Scarcity is a well-known problem in most of the developing and underdeveloped counties. In India, the scheduled communities constitute 15 per cent of the total population. Generally they are poor in economy and weak in awareness. Many of the members belonging to scheduled community work as daily wage earners in the fields. The inadequate educational status has been a reason for it. Members of scheduled communities have been suppressed for years together in India. Most of them still live in thatched houses and they have no permanent residence. They are economically very weak. For their financial requirements they borrow money from private money lender and paying higher rate of interest.

Hence, there is a need for deep study in Sathankulam Taluk, with the intention of know about the level of financial inclusion and financial knowledge among scheduled community.

10.3 Objectives of the Study

- To know the attentiveness about finance among Scheduled Communities in Sathankulam Taluk.
- To find out the saving habits of the people.
- To give suggestions to improve saving habits among Scheduled Communities in Sathankulam Taluk.

10.4 Hypothesis

- H_0 – No significant difference in financial awareness among members of scheduled communities by age, gender, marital status, religion and educational qualification.
- H_{02} – No significant difference in financial awareness among members of scheduled communities by occupation and monthly income.

10.5 Sampling Design

Under this study, primary data is collected by the researcher through convenience sampling method. Since the numbers of respondents are in numerous and the study period is limited, convenience sampling is employed. However, the 150 samples are selected from Sathakulam Taluk. Due representation is given for rural and urban respondents. Further, the respondents selected belong to various income groups.

10.6 Literature Review

- Review of Literature is important for every research to carry out the investigation effectively. Hence, the current study is based on the following reviews.
- Arockia Jerold, V (2008) in his technical paper on **"Financial inclusion and literacy in India"** had explained about the elements of financial inclusion, history of financial inclusion, measurement of financial exclusion, difficulties in promoting inclusive banking, strategies and approach towards financial inclusion, International experience in promoting financial inclusion, RBIs policy on financial inclusion, IT solution for financial inclusion, Role of government in promoting financial inclusion[1].
- Arun, T.J. and Ashok, J. (2010) in their study on, **"Financial Inclusion-Indian Experience"**, suggested that banks try to rearrange their business approaches to includedefinite plans to encourage financial inclusion of low income group. It is achance of business as well as a corporate social responsibility of every banks.
- AR Annadurai and P.PitchaiMuthu (2011) in their Article **"Financial Inclusion through financial literary and credit**

[1]Arun. T.J and Ashok. J, Financial Inclusion-Indian Experience, Indian Economic Panorama, 2010.

counselling centres" reveals that financial literary credit counselling centres will serve as hump for people especially from rural one as to achieve. Information pertaining to their economic activity. This study was a suggested that greater amount of financial education is required to the rural people to utilize the banking services. The banks are taking up various awareness programs relating to financial inclusion and financial literacy in their quoted villages while implementing the financial inclusion plan[2].

- Ammannya, k.k., (2007) in his technical paper on **"Financial inclusion-Tasks and strategies"** explained about financial inclusion in wide manner. It also explain about micro credit, its advantages and national challenges in achieving total financial inclusion.
- Sampath Kumara, V. (2008) in the study on, **"Technology for Financial Inclusion"** found that on the part of banking industry, extending the banking channels beyond branch services to excluded persons is a difficult venture. Digitalization has an important role to play in this regard. Appropriate technology with low operational cost network is the only way to extend banking services among all the branches. Technology has to allow the branches to go to the customer in place of the other way round.
- Murugadoss, K. and Ramesh, G. (2008) in their study on, **"Financial Exclusion and Financial Inclusion - An Introduction"**, suggested that financial service is essential for the betterment of every citizens of any country. The banking financial services closely related to the day today life of the human being of the world. The study identified that develop the banking service and access to the service, affordable credit facility and free face to face money advice to low income group.
- Selvaraju, R. and Vasanthi, G. (2008) in their study on, **"Rural Financial Inclusion"** suggested a new model for financial inclusion. Recognizing the need for providing social security, the model suggested in the study for rural financial inclusion includes no frills account, continuance of priority sector lending, micro finance, and micro finance products.

[2]Ammannaya, K.K. (2007), "Financial Inclusion Task and Strategies", Southern Economist, Vol. 45, No. 17, January 1.

10.7 Methodology

The principal data were composed from respondents with the support of interview schedule. The interview schedule was constructed by the researcher to collect information about financial literacy by visiting people in Sathankulam Taluk from 150 respondents has been selected by convenience sampling method.

10.8 Measurement of Financial Literacy

Financial literacy differs from individual to individual. Financial literacy does not purely dependent on educational literacy. Further financial literacy is a qualitative variable. Hence, literacy cannot be measured in absolute form. In order to measure the financial literacy, scaling technique has been employed. Scaling technique systematically converts the level of financial literacy of each individual respondent into its equivalent quantitative term. Financial literacy comprises of different elements. Hence to measure financial literacy of an individual in terms of different elements have been financially measured and then they are summed up together to get the final total financial literacy of an individual. Different elements of financial literacy identified by the researcher are:

- Financial literacy on investments.
- Financial literacy on borrowing.
- Financial literacy on debt management.
- Financial literacy on debit and credit cards.
- Financial literacy on capital Market.
- Financial literacy on insurance services.
- Financial literacy on postal services.
- Financial literacy on financial service.

From the above measurement, researcher has classified the schedule communities under two heads namely financially Literates and Illiterates.

10.9 Elements of Financial Literacy

Each questions included in the above eight elements carry one score. Altogether in eight elements of financial literacy, there are 50 questions. A scheduled community people with high financial literacy and answered positively for all 50 questions can get a maximum of 50 scores; and scheduled community people with low financial literacy can get a minimum score

Table 10.1 Elements of financial literacy.

S. no.	Element	No. of questions included
1	Investments	13
2	Borrowings	10
3	Debt Management	3
4	Debit Management	9
5	Capital Market	5
6	Insurance Service	3
7	Postal Services	3
8	Other Financial Services	4
	Total	**50**

of zero. Thus the level of financial knowledge among scheduled community people in Sathankulam Taluk is measured. Each respondent was asked to record their level of financial literacy with regard to these components.

Out of 50 questions, if the respondent answers any one of the questions correctly, he or she can become financially literate.

In order to measure financial literacy in relation to these eight elements, the above 50 questions were asked. And from the answers, the financial awareness level of the respondents in relation to the above elements is measured, and the details are furnished below.

10.10 Financial Literacy Among Scheduled Community

Financial literacy is necessary for everyone in the society especially for the scheduled community to access bank, postal, insurance and other

Table 10.2 Literacy scores.

S. no.	Financial literacy scores	No. of respondents	Percentage (%)
1	Below 1	38	25
2	1 to 5	96	64
3	5 to 10	16	11
	Total	**150**	**100**

financial services. The level of financial literacy among scheduled community people in Sathankulam Taluk is presented in table.

It is clear from the table that out of 150 respondents in the study 96 respondents (64 percent) who are financially literates have scored between 1 to 5 and 38 respondents (25 percent) who have the financial literacy scored below 1; and 16 respondents (11 percent) who have financial literacy scored 5 to 10.

10.11 Age Wise Status of Financial Literacy

Age is an important variable which has relationship with financial literacy. Age wise status of scheduled community in the study area is presented in Table 10.3.

Table 10.3 Age wise status of financial literacy.

Sl. no.	Status of financial literacy	Age group			
		20-30	30-40	Above 40	Total
1	Literates	38 (25.3)	26 (17.3)	48 (32)	112 (74.7)
2	Illiterates	10 (7.7)	7 (4.7)	21 (14)	38 (25.3)
Total		**48 (32)**	**33 (22)**	**69 (46)**	**150 (100)**

Source: Primary data

It is clear from the Table 10.3 that out of the total 150 respondents 48 respondents (32 percent) are financially literate in the age group of above 40; and 38 respondents (25.3 percent) are financially literates in the age group of 20 to 30; and 26 respondents (17.3 percent) are financially literates in the age group above the age of 40 is highly financially literate the other age groups.

10.12 Financial Literacy Among Scheduled Communities of Different Age Group - ANOVA

Members of the scheduled community at different ages have different level of financial literacy. As the age increase the knowledge on formal sources of finance and financial literacy than the younger ones. In order to study the relationship between financial literacy and age, the one way ANOVA is attempted with following null hypothesis.

10.12.1 Null Hypothesis

There is no significant difference in financial literacy among members of scheduled communities of different age groups in Sathankulam Taluk. The results of AVOVA are presented in Table 10.4.

Table 10.4 Financial literacy among scheduled communities of different age group – ANOVA.

Sl. no.	Financial literacy	Sum of squares	df	Mean squares	F	Sig
1	Between Groups	15. 752	2	7.876	1.681	.190
2	Within Groups	688.921	147	4.687		
Total		**704.673**	**149**			

Source: primary data

Table 10.4 reveals that in Sathankulam Taluk, since the 'p' value is higher than 0.05, the null hypothesis is accepted. It shows that there is no significant difference in financial literacy among scheduled communities and age wise in Sathankulam Taluk. It proves that age has not been a significant variable in financial literacy among scheduled communities in Sathankulam Taluk.

10.13 Financial Literacy and its Relationship with Gender

Gender is an important variable which has relationship with financial literacy. Generally female members of scheduled community are financially excluded and illiterates. In order to study the important differences in financial literacy among different gender groups, 't' test is endeavoured through the following null hypothesis.

10.13.1 Null Hypothesis

There is no important difference in financial literacy among members of scheduled communities of different gender group in Sathankulam Taluk. The results of 't' test are presented in Table 10.5.

It is clear from Table 10.5 that is no important difference in financial literacy among different gender group members belonging to scheduled

Table 10.5 Financial literacy among different gender group 't' test.

Gender	N	Mean	Std. deviation	Std. error of the mean	t test		
					T	Df	Sig
Male	81	3.31	2.10	0.23	1.607	148	0.110
Female	69	2.74	2.23	0.27			

communities in Sathankulam Taluk as evidenced through a higher 'p' value (0.110) at 5 per cent level of significance.

10.14 Financial Literacy and its Relationship with Marital Status

Both married and unmarried women are eligible to become financial literates. But among scheduled community, males are more financially knowledgeable than female. In order to study the relationship between financial literacy and marital status, 't' test is endeavoured with the following null hypothesis.

10.14.1 Null Hypothesis

There is no important difference in financial literacy among members of scheduled communities of different marital status in Sathankulam Taluk. The results of 't' test are offered in Table 10.6.

It is clear from Table 10.6 there is no important difference between financial literacy of different marital status belonging to scheduled communities in Sathankulam Taluk as evidenced through a higher 'p' value (0.056) at 5 per cent level of significance.

Table 10.6 Financial literacy among different marital Status – 't' test.

Marital status	N	Mean	Std. deviation	Std. error of the mean	t test		
					T	Df	Sig
Male	128	2.91	2.20	0.23	1.925	148	0.056
Female	22	3.86	1.88	0.27			

10.15 Financial Literacy and its Relationship with Religion

Scheduled community in Sathankulam Taluk follow two types of religion such as Hinduism and Christianity. Religion has also relationship with financial knowledge. In order to study the association between financial knowledge and religion, 't' test is endeavoured with the following null hypothesis.

10.15.1 Null Hypothesis

There is no important difference in financial literacy among members of scheduled communities of different religion in Sathankulam Taluk. The test of ANOVA outcomes are presented in Table 10.7.

It is clear from Table 10.7 that there is no important difference in financial knowledge among members of different religions belonging to scheduled communities in Sathankulam Taluk as evidenced through a higher 'p' value (0.951) at 5 per cent level of significance.

10.16 Financial Literacy Among Scheduled Communities of Different Educational Qualification – ANOVA

Education is the basis for financial development. Education develops every individual to live a meaningful and standard life. It is vital for rapid development. Most of the scheduled community in Sathankulam Taluk is illiterate. In order to study the association between monetary literacy and educational qualification, the one way ANOVA is endeavoured with the following null hypothesis.

Table 10.7 Financial literacy and its relationship with religion.

Religion	N	Mean	Std. deviation	Std. error of the mean	t test		
					T	Df	Sig
Hindu	90	3.06	2.18	0.23	0.061	148	0.951
Christian	60	3.03	2.19	0.28			

Table 10.8 Financial literacy among scheduled communities of different educational qualification – ANOVA.

Financial literacy	Sum of squares	Df	Mean squares	F	Sig
Between Groups	22.811	5	4.562	.963	.442
Within Groups	681.862	144	4.735		
Total	**704.673**	**149**			

10.16.1 Null Hypothesis

There is no importantvariance in financial literacy among members of scheduled communities of different educational qualification in Sathankulam Taluk. The tests of ANOVA result are presented in Table 10.8.

Table 10.8 reveals that in Sathankulam Taluk, since the 'p' value is greater than 0.05, the null hypothesis is established. It shows that educational qualification wise there is no importantvariance in financial literacy among scheduled communities in Sathankulam Taluk.

10.17 Occupation Wise Status of Financial Literacy

Occupation wise status of financial **literacy** among members of scheduled community in the study area is presented in Table 10.9.

Table 10.9 shows that among the financially literate respondents 70 respondents (47 %) are coolie workers and 14 respondents (9.3 %) are unemployed and 8 respondents (5.2 %) are businessman and 7 respondents (4.7 %) are farmers and 7 respondents (4.7 %) are government employees and 6 respondents (4 %) are working in private organization.

10.18 Financial Literacy Among Scheduled Communities of Different Occupation ANOVA

Members of scheduled community are doing different activities. They have different occupation such as coolie, agriculture, and business, government job, professional and working in private organization. In order to study the association between financial **literacy** and occupation, the one way ANOVA is attempted with the following null hypothesis.

Table 10.9 Occupation wise status of financial literacy.

Sl. no.	**Status of financial literacy**	**Educational qualification**						
		Coolie	**Farmer**	**Business**	**Govt employee**	**Employee in pvt organization**	**Un employed**	**Total**
1	Literates	70 (47)	7 (4.6)	8 (5)	7 (4.6)	6 (4)	14 (9.3)	112 (75)
2	Illiterates	27 (18)	1 (0.6)	3 (2)	1 (10.6)	3 (2)	3 (2)	38 (25)
	Total	**97 (65)**	**8 (5.2)**	**11 (7.3)**	**8 (5.2)**	**9 (6)**	**17 (11.3)**	**150 (100)**

Table 10.10 Financial literacy among scheduled communities of different occupation ANOVA.

Financial literacy	Sum of squares	Df	Mean squares	F	Sig
Between groups	5.367	5	1.073		
Within groups	699.306	144	4.856	.221	.953
Total	**704.673**	**149**			

10.18.1 Null Hypothesis

There is no important variance in monetary **literacy** among members of scheduled communities of different occupation in Sathankulam Taluk. The test of ANOVA results are presented in Table 10.10.

Table 10.10 reveals that in Sathankulam Taluk, since 'P' value isgreater than 0.05, the null hypothesis is established. It shows that Occupation wise there is no important difference in financial literacy among scheduled communities in Sathankulam Taluk.

10.19 Monthly Income Wise Status of Financial Literacy

Monthly income is also one of the important variable which has an inter relationship with financial awareness. Monthly income wise statuses of financial awareness among the members of the scheduled community in the study are is presented in Table 10.11.

Table 10.11 Monthly income wise status of financial literacy.

Sl. no.	Status of financial literacy	Monthly income (Rs) Below 2000	2000 - 4000	4000 - 6000	Above 6000	Total
1	Literates	3 (2)	40 (26.7)	54 (36)	15 (10)	112 (74.7)
2	Illiterates	4 (2.7)	11 (7.3)	18 (12)	5 (3.3)	38 (25.3)
	Total	**7 (4.7)**	**51 (34)**	**72 (48)**	**20 (13.3)**	**150 (100)**

It is clear from the Table 10.11, 54 respondents (36 %) earning a monthly income of Rs.4000 to 6000 have become financially literate and 40 respondents (26.7 %) earning a monthly income of Rs.2000 to 4000 have become financially literate and 15 respondents (10 %) belongs to the earning category of above 6000 have become financially literate and 3 respondents (2 %) belong to the earning category of below Rs.2000 have become financially literate.

10.20 Financial Literacy Among Scheduled Communities of Different Monthly Income - ANOVA

Members of scheduled communities have different levels of income. The aim of the financial awareness is to improve the economical setup of the people who are living below the line of poverty. In order to study the association between monetary awareness and monthly income, the one way ANOVA is endeavoured with the following null hypothesis.

10.20.1 Null Hypothesis

There is no important difference in financial awareness among members of scheduled communities of different monthly income in Sathankulam Taluk. The test of ANOVA results are presented in Table 10.12.

Table 10.12 reveals that in Sathankulam Taluk, since the 'p' value is bigger than 0.05, the null hypothesis is accepted. It shows that monthly income wise there is no importantvariance in financial awareness among scheduled communities in Sathankulam Taluk.

Table 10.12 Financial awareness among scheduled communities of different monthly income - ANOVA.

Sl. no.	Financial literacy	Sum of squares	Df	Mean squares	F	Sig
1	Between groups	3.915	3	1.305		
2	Within groups	700.758	146	4.800	.272	.846
Total		**704.673**	**149**			

10.21 Conclusion

In Sathankulam Taluk the members of scheduled community are literates at a lower level. This is because of their lack of education on financial literacy. These lack of literacy can be removed through advanced wireless mobile network.

The Government may takes many steps to recover the level of monetary inclusion and monetary literacy among scheduled communities through advanced wireless network during this pandemic period. To recoverthe financial literacy level, Government must take steps like awareness programmes and digital financial literacy programmes among scheduled communities.

10.22 Bibliography

- Atkinson, A., & Messy, F. (2012). Measuring financial literacy: Results of the OECD/International Network on Financial Education (INFE) pilot study (OECD Working Papers on Finance, Insurance and Private Pensions, No. 15). Paris: OECD Publishing. DOI: http://dx.doi.org/10.1787/5k9csfs90fr4-en.
- Kiliyanni, A.L., &Sivaraman, S. (2016). The perception – reality gap in financial literacy: Evidence form the most literate state in India. International Review of Economics Education, 23, 47 – 64.
- Jerold Arockia, V. (2008), "Financial Inclusion and literacy in India," Indian Journal of Finance, Vol.2, N0. 3, July.
- M. Ramji, Financial Inclusion in Gulbarga: Finding Usage in Access, Chennai; IFMR, 2009.
- Mitra Rona (2006), "Financial Inclusion: Meeting the challenge", people's Democracy, April 22 & 29, 2007.
- UshaThorat (2007), 'Financial Inclusion, The Indian Experience', Text of the speech delivered at the H MT-DFID Financial Inclusion Conference White hall place, London, June 19, 2007, RBI, Mumbai.
- Arun. T.J and Ashok. J, Financial Inclusion-Indian Experience, Indian Economic Panorama, 2010.
- Ammannaya, K.K. (2007), "Financial Inclusion Task and Strategies", Southern Economist, Vol. 45, No. 17, January 1.

11

Advanced 5G Technology in Healthcare Field

R. Delshi Howsalya Devi[1], R. Naresh[2*] and C.N.S.Vinoth Kumar[2]

[1]Department of Artificial Intelligence and Data Science, Karpaga Vinaya College of Engineering & Technology, Chengalpattu, Chennai, Tamil Nadu, India
[2]Department of Networking and Communications, SRM Institute of Science and Technology, Kattankulathur, Chengalpattu, Chennai, Tamil Nadu, India

Abstract

Our everyday lives are already more integrated with wearable technology, and soon our bodies will follow suit. The inevitable, substantial rise in the population of older persons has increased the need for personal-health monitoring and illness prevention. The cost of detecting, preventing illnesses, and saving patient lives can be significantly decreased by the technologies with 5G in wearables and healthcare. In this research, 5G technologies for wearable technology and healthcare are reviewed. Wearables and 5G might be cutting-edge healthcare solutions. This study examined the advantages of 5G technologies, which are used in wearable technology and healthcare. It has the ability to directly influence clinical judgement. The monitoring of ongoing human physical activity and patient rehabilitation outside of hospitals might both be facilitated by this technology. This article divided the roles into the different categories to review the survey such as, Patient health monitoring using 5G, Continuous monitoring of chronic diseases using 5G, Management of preventing infectious diseases using 5G, Robotic surgery using 5G and 5G and the future of wearables. Due to the adoption of 5G technology by healthcare systems, sick people often receive adequate treatment easily and have a connection with professionals who are otherwise unavailable. Thanks to 5G, healthcare systems now have the ability to allow mobile networks to host telemedicine meetings, expanding the reach of the telemedicine programme.

Objectives

Our everyday lives are already more integrated with wearable technology, and soon our bodies will follow suit. The inevitable, substantial rise in the population

**Corresponding author*: nareshr@srmist.edu.in

A. Suresh, J. Ramkumar, M. Baskar and Ali Kashif Bashir (eds.) Resource Management in Advanced Wireless Networks, (227–246) © 2023 Scrivener Publishing LLC

of older persons has increased the need for personal-health monitoring and illness prevention. The cost of detecting, preventing illnesses, and saving patient lives can be significantly decreased by the technologies with 5G in wearables and healthcare. In this research, 5G technologies for wearable technology and healthcare are reviewed.

Methods

Wearables and 5G might be cutting-edge healthcare solutions. This study examined the advantages of 5G technologies, which are used in wearable technology and healthcare. It has the ability to directly influence clinical judgement. The monitoring of ongoing human physical activity and patient rehabilitation outside of hospitals might both be facilitated by this technology.

Results

This article divided the roles into the following categories to review the survey results:

Patient health monitoring using 5G

The distribution paradigm for wearables for patient monitoring and healthcare is quickly shifting from a patient-centric approach to a conventional hospital and specialist-focused one. As individuals are now expected to travel to a clinic or hospital for medical care, the health care system is plagued with problems. A possible solution for addressing this issue is 5G health tracking. Smartphone using the 5G wireless mobile network, the next-generation wireless mobile network. It's a 5G-capable in-home health monitoring equipment that will put extra strain on the networks of the healthcare system. In particular for healthcare practitioners who may be in communication with tens of thousands of patients every day, this results in slower network speeds and more congestion. Poor quality can lead to patient care delays in addition to the time delay, thereby compromising long-term outcomes. Since they can keep patients and doctors in constant communication, 5G technologies have the ability to address these issues while still reducing total medical expenses.

Continuous monitoring of chronic diseases using 5G

Continuous monitoring improves chronic patients' quality of life while decreasing the expenses of the sanitary system. The results demonstrate that the 5G network is required to continually follow a large number of patients since it offers reduced latency and guarantees enough bandwidth for all users. Hence, remote patient nursing and tracking capabilities, as well as advancements in data connectivity between patients and health-care providers, may signify a paradigm change in 5G technologies. The healthcare technologies that 5G enables include telemedicine, robotic surgery, and augmented reality/virtual reality aided surgery, to name a few.

Management of preventing infectious diseases using 5G

Healthcare-related infections are becoming more common these days. A recent advancement in healthcare is the use of 5G to manage these infectious diseases. According to projections, 5G networks may support up to one million wired devices per square kilometre. Wearable technology can therefore quickly stop the transmission of illness among those who are ill.

Robotic surgery using 5G

Remote robot assisted surgery is being performed by doctors utilising 5G. The surgeon performed the procedure by controlling robotic invasive arms over a 5G network using robotic surgery. Its quick pace will lessen the possibility of possibly deadly medical mistakes. It was around 1,500 kilometres to the patient. It can lower operation expenses while raising the level of treatment provided to patients in remote locations.

5G and the future of wearables

With 5G, wearable devices like smartwatches and fitness trackers may connect people to the outside world in whole new ways. On 5G networks, which are expected to support up to one million connected devices per square kilometre, there should be enough capacity to accommodate athletes and spectators without congestion or device slowdowns. As sensor technology develops, health trackers could be able to gather high-quality patient data and provide doctor-driven diagnoses, enhancing the influence of 5G on healthcare.

Conclusions

The recent exponential expansion of wearable technology has increased study interest in a variety of fields, including as communications and healthcare. Due to the adoption of 5G technology by healthcare systems, sick people often receive adequate treatment easily and have a connection with professionals who are otherwise unavailable. Thanks to 5G, healthcare systems now have the ability to allow mobile networks to host telemedicine meetings, expanding the reach of the telemedicine programme.

Keywords: 5G, wearable devices, healthcare, chronic disease, health monitoring, telemedicine

11.1 Introduction

5G will replace the present 4G communication network and bring new features including scalability, improved capacity, and speed. The 5G concept's standards, capabilities, and technologies are presently being thought about and discussed. It is anticipated that 5G will have a big impact on the advancement of the next-generation of intelligent healthcare. The term "5G" refers to the fifth generation of wireless transmission technology, which is anticipated to have a significant influence on many aspects of contemporary society, including wearable technology and healthcare. Presently, for applications in smart healthcare. The current 4G network is widely used, as are other forms of communication. Also, they are always adapting to the needs of applications for intelligent healthcare in the future. Current communication technologies are unable to meet the complex and dynamic demands that smart healthcare applications make on

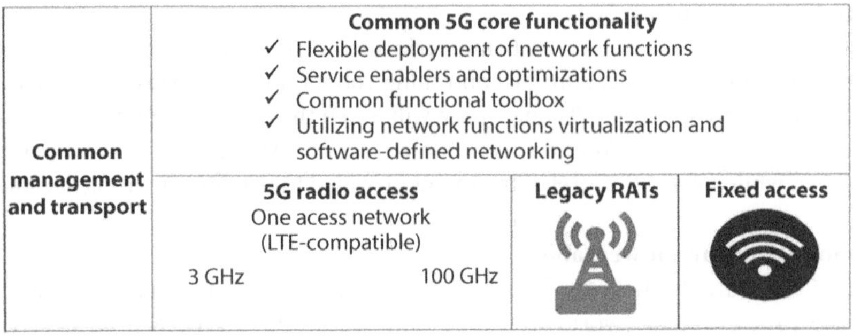

Figure 11.1 High-level 5G architecture.

communication networks. Wearables will get real-time notifications, more precise measurements, and more information about our surroundings, wellbeing, and objectives thanks to 5G.

The forthcoming 5G network should therefore be able to support smart healthcare applications that satisfy the majority of requirements, including high density, ultra-high reliability, high energy efficiency, bandwidth, and ultra-low latency. Smart healthcare is one of the most important uses for 5G networks. This article provides a thorough review of 5G-enabled wearables and smart healthcare. The 5G architecture is shown in Figure 11.1 [1].

11.2 Objectives

- Users will be able to connect to the globe in whole new ways with 5G.
- It makes it possible to do complicated tasks close to the user.
- Smarter, more streamlined, and more potent wearables that can utilise other 5G-enabled equipment and have longer battery lives.
- Real-time, secure communication between medical professionals and patients is possible with 5G.
- 5G-enabled wearables increase effectiveness, enabling professionals to visit more patients.
- Able to regularly add more heavy features to their wearable gadget, compute more, and require less charge.
- Running routes can be tracked using wearables with 5G capabilities.
- 5G-capable wearables can keep an eye on vital indications.

11.3 Our Contribution

Many studies on 5G technology in healthcare and wearable devices have been conducted. This section provides an overview of a few 5G technologies that are being used in wearable devices and healthcare.

11.3.1 Patient Health Monitoring Using 5G

A wireless Body Area Network has been used to monitor patients remotely in a number of studies. Before delivering the movie to the doctor, the idea is to review a variety of bright indicators restrictions recorded by various sensors placed on the body exterior, or even fixed sensors, and to capture all motions using a wearable receiver or radiocommunication gateway. Clinicians need the automatically collected data in order to correctly diagnose the patient's actual condition. The inclusion of the word "responsible" in the title of this article is not intended to imply any kind of endorsement by the U.S. government.

New possibilities for action control, patient monitoring, and data analysis might be made possible by 5G network technologies. For instance, the prevalence of cardiac illness would soar as a result of a future 5G network's greater communication capacity, and patient follow-up and medication administration will be provided in novel methods. Another crucial use would be an improved method for modifying medicine according to individual needs. As an example, insulin dosages for diabetic patients are now given in doses based on past experience and maybe regular interval-based monitoring. cardiac patients' quality of life will improve if implanted dose-reservoirs can be managed remotely. High-bandwidth health care professionals will be able to connect with one other, a central server, and patients via a Personal Trusted Gateway thanks to 5G (PTG). Some academics have investigated using 5G technology to monitor patients. The report that follows will give a brief rundown of recent 5G research for patient health monitoring.

A paper titled "Real-Time Remote Health Monitoring System Powered by 5G MEC-IoT" was submitted by Yangan Zhang and colleagues [20]. They launched An artificial intelligence and MEC-based telemedicine system has been created for remote illness diagnosis and health monitoring. Their research suggests that combining diverse technologies, including computers, medicine, and telecommunications, can significantly improve patient treatment effectiveness while reducing healthcare expenditures. In simulation findings, their recommended model exhibits impressive prediction accuracy over a wide range of categories in the ECG dataset, enabling the entire system to communicate medical information more effectively.

Lalita *et al.* [19] suggested a Based on the 5G NR (New Radio) architecture, a four-layer design for an ehealthcare system that integrates the control plane and user plane. They assessed the throughput and latency for different values of the OFDM numerologies while running the simulation throughout the two frequency ranges (1 and 2). They tested 4G and 5G, and found that 5G has a 10 times lower latency than 4G and can accommodate a far greater number of devices. With 5G NR, electronic communication is also delivered for better healthcare facilities. Discussion completes data transmission and speeds up and simplifies illness diagnosis.

Abdul and colleagues [17] provide a current assessment of the state-of-the-art of smart healthcare made possible by 5G and IoT, encompassing research trends, challenges, taxonomy, and potential directions for future study. They started by presenting a 5G smart healthcare architecture and the methods required to make it work. Second, they discussed the new requirements and goals for 5G smart health care and provided the taxonomy for that field. They provided a thorough review of the network layer solutions for IoT-based 5G smart congestion control, healthcare, including routing, and scheduling that took into account both existing work and prospective future research in the third stage. Lastly, they talked about some of the issues and difficulties that may affect 5G smart healthcare in the future.

Lakhal *et al.* [18] suggest creating a wireless sensor network application that uses 5G technology for high-speed transmission to increase the safety of healthcare personnel. It entails sending medical data to a hospital server via the Internet using a variety of sensors attached to the patient's body. These sensors assess physiological signs in real time and deliver medical data that may be used for monitoring and diagnosing.

The doctor will therefore be able to obtain and review the patient's medical data via a JEE-designed communication interface that can receive an SMS in the event of an emergency. A smartphone is utilised by the patient's side to process the data acquired by the wearable devices, which are used to obtain measurements from the patient's body. To discover and create alarms, this blends machine learning in Big Data from several hospitals with patient data. To evaluate the practicality of 5G in our design, research experiments that simulated traffic flow from a lot of users to the Data Base were carried out. The research also showed that the 5G network should be used to continuously monitor a large number of patients since it has small latency and guarantees bandwidth availability for all users.

Continuous Health Status Monitoring by 24x7 Sensing and Transmission of Physiological Data via 5-G Cellular Channels was developed by A. Mishra and D. P. Agrawal [15]. They selected the electrocardiogram and pulmonary artery pressure as the early physiological signals for this activity because they can aid in the diagnosis of cardiovascular diseases, which are the most frequent silent killers in the modern day. It utilised a Wireless Body Area Sensor Network to gather physiological data from a user's body and transmit it to a WBAN coordinator. This technology is distinctive in that it enables cellular system subscribers to have an online, 24/7 health monitor with extremely little power usage by collaborating with the underlying sensor networks.

11.3.2 Continuous Monitoring of Chronic Diseases Using 5G

Most chronic illnesses require monitoring to determine if the disease is advancing or regressing as well as the appearance of consequences. Making these decisions about what to measure, when to track it, and how to modify therapy is necessary. A loss of control, ineffective use of time, and potentially detrimental treatment adjustments can result from poor judgements in any of these areas. To aid in the management of a chronic or recurrent ailment, monitoring is a sort of measurement that is performed often. Patients, clinicians, or both may take part. Monitoring and adjustment can be managed by clinicians or patients [11–15].

Continuous mobile health monitoring systems that deliver medical assistance across long distances enhance people's quality of life. This technique enables a patient to speak with a doctor across a great distance if he falls unwell. It reduces the time and expense required to obtain medical care. A 5G network and machine-to-machine communications are integrated in order to establish and create continuous monitoring of chronic illness (M2M). A significant advancement in M2M communications and IoT is required to improve healthcare solutions. characteristics that could be anticipated in 5G technology.

- Peak data rates should vary from tens to hundreds of gigabits per second.
- Very low latency (as low as 1 ms).

- In terms of energy utilisation, the network will be 100 times more efficient. In terms of spectrum use, the network will be three times more effective.

The continuous monitoring system is composed of a number of wearable devices, a smartphone, and a database server. In this suggested assessment, the smartphone is connected to the server using 5G technology, and Figure 11.2 shows the architecture overview. Many physiological variables, including blood pressure, heart rate, blood sugar, and others are continually monitored by the wearable sensors, which are Bluetooth-connected to the smartphone. A 5G connection to the server is present on each smartphone.

Singh *et al.*'s [14] Smart M-Health Continuous Monitoring System Utilizing 5G Technology includes a smart phone, wearable sensors and gadgets, and a database system that stores all patient data. The status of a patient is continually predicted by artificial intelligence, which also determines whether or not the patient requires medical assistance.

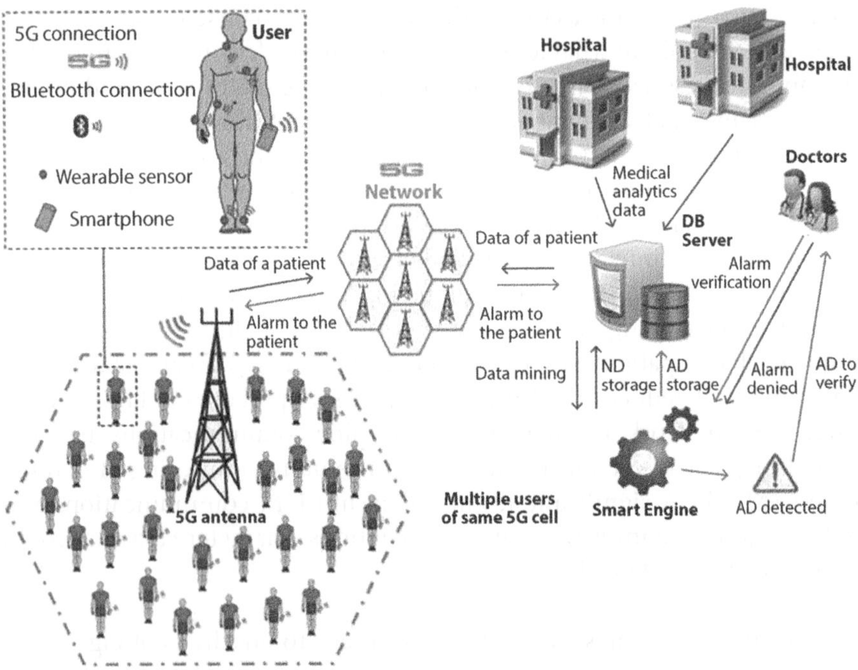

Figure 11.2 5G network for smart continuous monitoring system [21].

This technique enables a patient to speak with a doctor across a great distance if he falls unwell. Medical aid takes less time and costs less money. Index Terms: Machine to Machine (M2M), Artificial Intelligence (AI), Abnormal Data (AD), and M-Health (AI).

Oleshchuk and Fensli [13] examined potential outcomes in the context of 5G communications networks as well as associated security and privacy issues. One of the novel issues in comparison to the present is the potential for direct access to devices within the BAN. They advised us to establish a balance in medicine between patient privacy and confidentiality, the need for usefulness and safety, as well as strategies for handling emergency circumstances. On the other hand, the existing systems for authentication and authorisation were not created with these conditions in mind. Based on this scenario, they came up with a unique method for handling emergency circumstances that could occur in the medical setting by expanding the RBAC model.

Sigwele *et al.* [12] presented 5GREM, a remote monitoring system for cardiovascular illness that employs an ECG biosensor as a BSN device. In order to avoid heart attacks and heart failure by effective vigour, its objective is to check and analyse the patient's cardiac beats and to send emergency alerts to nearby ambulances, caretakers, or doctors during disruptions. Demands from the ECG sensor are either shifted nearby on the entry, transferred to nearby mobile devices, or routed to the 5G authority depending on the battery level, CPU level, transmission power, delays, and task failure rate. The 5GREM framework makes advantage of D2D task offloading, Mobile Edge Computing (MEC), and 5G technology to minimise energy usage in the mobile gateway. According to 5GREM, heart attack fatalities can be significantly decreased.

According to Frost & Sullivan [22], 5G combined with machine intelligence, data analytics, and the Internet of Things (IoT) could enable new ways to tailor diagnostic tests while lowering hospital visits and medical expenditures. They looked at smartphone-based saliva testing and blood glucose self-monitoring. Driving the Digital POCT Market, a report, demonstrates the growing popularity of digital point-of-care testing (POCT) solutions. More individuals choose self-monitoring for fast diagnosis of respiratory illnesses and saliva sample-based diagnostics. The digital POCT industry, which covers remote testing requirements, is being driven by the COVID-19 epidemic and growth in diabetic patients. New methods to customise diagnostic tests while reducing hospital stays and medical expenses are on the horizon with the introduction of 5G, machine-based intelligence, data analytics, and the internet of things (IoT).

11.3.3 Management of Preventing Infectious Diseases Using 5G

IoT and 5G communication are intended to increase communication networks to enhance people's quality of life from a social and humanitarian perspective. Technology-based control methods may prove to be an effective weapon in the battle against the spread of infectious diseases. This study examines the potential applications of cutting-edge technology, such 5G, for controlling the spread of infections, managing infectious disease prevention, and guaranteeing patient safety. The healthcare industry can provide a consistent, rapid, and continuous service to people throughout the spread of disease by utilising modern wireless networking technologies like 5G. In terms of improved dependability, connection stability, ultra-massive accessibility, network scalability, and rapid response flexibility, it may help support the strong communication infrastructure of the smart healthcare system, allowing pandemic monitoring and prevention.

The enhancement of the health, medical, and socioeconomic sectors can all be facilitated by wireless communication technology. Hospitals in disaster circumstances are soon overrun by the enormous number of patients in need of care. Controlling the problem is difficult since there aren't enough medical staff and resources, and there aren't enough beds. In fact, because of the virus's ability for fast transmission, conditions actually get worse in pandemic circumstances. Avoiding interaction between a patient seeking therapy and the doctor is an excellent way to get around this problem.

In their analysis of the technology-driven, 5G-enabled e-healthcare system during the COVID-19 pandemic, Nasser Alshammari *et al.* [11] identified eight key areas of application for COVID-19 control, including illness detection, travel history analysis, symptom identification, early diagnosis, transmission identification, information access during lockdown, the development of medical treatments and vaccines, as well as human mobility. It shows that the majority of individuals can easily get information online, on social networking sites, and on television. The results showed that 42% of respondents reported feeling tense all or most of the time during the COVID-19 pandemic. Just 28.6% of responders to the COVID-19 issue indicated they were always nervous, while the remaining 30% said they were never uncomfortable.

11.3.4 Robotic Surgery Using 5G

The robot technique has just recently been used in clinical practise in the orthopaedic profession. It improves the accuracy and reproducibility of implant placement, and it has a lot of promise to improve and provide a better and safer clinical outcome for orthopaedic surgery. The remote surgical idea is based on the exchange of medical data. Digitalized medical information is delivered across cable or wireless communications networks, including images, audio, and video. The networks can be used by surgeons to operate and control surgical robots remotely [2–5].

The purpose of this study was to explore if employing a robot would assist demonstrate the efficacy and feasibility of 5th generation wireless technology (5G). Surgeons may operate a medical robot or other surgical tool remotely via 5G connection, just as if they were in the same room. With the use of 5G technology, several professionals will be able to collaborate for the first time to perform cutting-edge surgery from anywhere in the globe in just a short amount of time. The first 5G remote brain operation has been finished by medical professionals. The Parkinson's patient was around 1,500 kilometres away and had the condition. With a three-hour procedure, the patient recognised a deep brain stimulation implant. Dr. Ling Zhipei carried out the ground breach operation from his home in Sanya City, controlling equipment in Beijing with a computer powered by China Mobile and Huawei's 5G network.

The 5G networks might be useful for robotic surgery care. With 5G networks, video conversations, gaming, and download speeds will all improve. On the other hand, specialised doctors predict an increase in long-distance, robotically aided surgery. The advantages of a 5G connection are beneficial for both telepresence and telesurgery. On the other hand, the promise of 5G extends beyond merely improving the speed of today's technology. The higher number of data that can be analysed quickly will enable surgeons to combine medical equipment and collect data in previously unthinkable ways. For the first time, specialists from around the globe could collaborate. Long-term, it may save operational costs while simultaneously improving the standard of care provided to patients in remote locations.

The two primary forms of robotic surgery are assistive and completely automated. In order to ensure that medical devices are properly positioned, seal up patients after surgery, and make tiny incisions, doctors employ assistive technology. Without human assistance, fully automated surgical equipment can complete whole surgeries. According to the IEEE, work on fully autonomous surgical robots is now occurring. A pig's intestine was successfully repaired by the Smart Tissue Autonomous Robot (STAR) in 2016 with a lot more accuracy than a human surgeon.

They carried out the first 5G surgery animal testing ever, along with Huawei, China Unicom Fujian Branch Mengchao Hepatobiliary Hospital, and Suzhou Kangduo Robot Co., Ltd. With very little downtime, the treatment is finished in about 60 minutes. The surgical incisions were clean, and no blood was spilled throughout the treatment. The vital signs of the test animals did not change following surgery. His operation was successful in achieving this objective because to the massive bandwidth, extremely low latency, and robust connection capabilities of Huawei's 5G net technology. High-definition video has achieved the same experience as a fiber-optic line based on 5G network control expertise, handling both the control connection and the two video connections at both tops of the remote-control machine. The doctor provided a glowing recommendation. When 5G medical association applications advance, remote diagnosis, remote treatment, and remote surgery will progressively be deployed, significantly extending the lifespan of the sick.

Tian *et al.* [4] suggested Telerobotic Spinal Surgery Utilizing a 5G Network. They wanted to know if our first 12 patients would respond well to telerobotic spinal surgery using 5th generation wireless networks (5G). They came to the conclusion that 5G remote robot-assisted spinal surgery is precise and trustworthy, and 5G telerobotic spinal surgery is both effective and practical for the safe management of spinal illnesses. They also came to the conclusion that telerobotic spinal surgery based on 5G technology is precise, secure, and reliable. A more recent technique with a lot of potential is spine surgery robotics. With robotically inserted devices, surgical accuracy seems to be quite good. On the other hand, the impact of robotics on radiation exposure is uncertain and seems to rely on the strategy and kind of robot used.

11.3.5 5G and the Future of Wearables

In the upcoming years, wearables' capabilities may go beyond health and wellness, offering new levels of value to apps that call for a distant uplink thanks to 5G's ultra-low latency. An excellent substitute for senior live-in care might be ambient assisted living made possible by 5G-enabled wearables. A 5G-enabled wearable that may notify loved ones or medical professionals of a drop in blood pressure or an untaken dose of medication might improve the quality of life for patients and carers overall. Wearables may potentially offer more specialised experiences in smart homes. In addition to learning thermostats and smart locks, 5G promises to free developers from speed and compatibility restrictions, enabling customers to experience more seamless and intuitive connectivity. Also, it may open the door for the subsequent AI-powered virtual assistant, one that attends to every member of the family and provides what is needed without being asked. With the advent of 5G, wearables might go from being merely convenient additions to vital tools that teach users more about their environment and themselves. Verizon's 5G Ultra-Wideband has brought that future closer.

Speed, though, isn't the most alluring feature of 5G for wearables. To begin with, 5G offers far lower latency than 4G. The time it takes for data to travel between two locations drops from around 20 milliseconds to about 1 millisecond. That's nearly instantaneous in the majority of usage cases, which is essential for an autonomous vehicle to determine if its path has to change immediately away. Major 5G interruptions may be categorised into the following important categories: Device-centric architectures.

- Wearable antennas
- Massive multiple input/multiple output (MIMO)
- Smart devices and the IoT
- The mmW band

11.3.5.1 Internet of Things in 5G Communications

More than any other new collection of wireless technologies, the Internet of Things (IoT) and 5G are revolutionary. This is due to the fact that 5G networks would suggestively increase the performance and dependability of these connected gadgets. Many industries will see growth thanks to 5G, and it will provide a platform for emerging technologies like the Internet of Things to become ingrained in our culture and way of life. 5G is the cornerstone for addressing the Internet of Things' full potential. The majority of operator IoT revenue now comes from connection, but in the next five years, money will also come from apps, services, and platforms that allow services. In order to connect and protect next-generation devices, such as Cinterion IoT Modules and IoT projects as well as 5G SIMs, Thales offers a wide choice of 5G solutions, IoT gateways, and modem cards. These solutions also make it simple to shift to new networks and structures.

Beyond an increase in speed, 5G networks will be more reliable, leading to more reliable connections. Having a trustworthy and steady network state is crucial for any IoT, but especially for connected devices like security cameras, locks, and other monitoring systems that depend on real-time updates. The next-generation network must have very low latency, high-speed connection, and extended coverage for IoT devices. To benefit from these developments, manufacturers must first engage in 5G-compatible goods. For IoT-enabled devices, 5G connects more devices at faster speeds and almost eliminates latency. The term "mobile IoT" refers to cellular low-power wide-area systems that operate on authorised frequency bands. Both 3GPP Narrowband IoT and Long-Term Evolution Machine-Type

Communications technologies will be used in the next 5G era of intelligent communications. Nowadays, 5G may be used to access mobile IoT solutions for smart logistics, smart utilities, and smart cities. The earliest 5G applications and customer premises equipment included fixed wireless access, mobile computing, video broadcasting, and enterprise and high-speed industrial networking.

- With 5G IoT, facilities will be able to send important upgrades to whole networks without having to halt operations, overwhelm servers, or cause functionality to freeze.
- As personal apps are fundamentally altering how we work and live, 5G IoT will raise the standard of living for regular people.
- The following current sectors will continue to benefit from 5G IoT developments:
  - Smart Utilities
  - Agriculture
  - Smart Cities
  - Security and Surveillance
  - Smart Buildings
  - Healthcare
  - Smart Factories
  - Automotive and Transportation

11.3.5.2 5G and Business IoT

5G-enabled The Internet of Things is anticipated to not only speed up technical development but also support 22 million jobs worldwide. This job growth is probably being driven by the digitalization of the transportation, industrial, agricultural, and other physical industries. The productions at mines, oil rigs, construction sites, and freighter fleets would benefit greatly from ultra-fast data transport due to the nature of their time-sensitive output.

11.4 Challenges Facing Wearable Technology

Like any new technology, wearable computing devices go through an introduction and exploration phase. A new trigger that will quicken the life cycle of wearable technology is the coming of 5G, which will offer greater bandwidth and open the door to many options like real-time health monitoring and several apps running at once. What uses these wearables and 5G combo will be useful for is easy to conceive. Some customers will easily dismiss the devices, while others will periodically doubt their value and purpose. On the other hand, wearable technologies have much too much potential and provide far too many benefits to be ignored. The future of wearable computing devices will be equally as bright, with only our imaginations acting as a limit, if the adoption of smartphones is any indication.

- Improvements in software architecture to make up for the challenges of using tiny screens for navigation.
- Wireless and local area network administration.
- Adequate protection from hackers who may obtain access to the device's data, if not the actual device, while it's in use.
- The use of autonomous processing and augmented reality has become dependent on technology.

11.5 Benefits of 5G

- Verizon 5G and the Phoenix Suns are revolutionising sports.
- More gamers are using the cloud as a result of 5G.
- Betting on the financial and 5G futures.
- 5G and the decentralisation of the film industry.
- Upcoming Uses for the 5G Network.
- Near-real-time remote collaboration is possible with 5G network slicing.5G small business potential.
- Ghost Kitchens: How 5G can transform the food delivery industry, one order at a time.
- Water and Technology: Updating Our Water Infrastructure With 5G and IoT.
- 5G delivers cutting-edge entertainment, including robot dogs and 3D holographic displays.

11.6 Next Generation of Healthcare with the IoMT and 5G

A significant subsection that includes everything related to health care is the Internet of Medical Things (IoMT). Particularly 5G has the ability to dramatically broaden the IoMT frontier. The creators of these devices may now take use of the 5G data transfer speeds necessary for linking IoMT devices to each other, to the successful servers and databases that keep track of them, as well as for real-time solutions for surgeons. These tools may help save the time a patient spends in the hospital while also keeping physicians more regularly connected to their patients, which might theoretically lead to fewer hospital visits and better overall care.

The use of IoMT technology has the potential to significantly impact two key facets of healthcare: hospital procedures and expenses, as well as home care. We all receive a plastic wristband when we are admitted to the hospital, whether it is for a reserve or a standard procedure. It includes private data like our name, blood type, birthday, and so on. On the other hand, your subsequent bracelet can resemble a smartwatch more. It may be connected to information about your health, indicate where you are in the hospital, or even monitor your vital signs. This bracelet may also be connected to a private 5G network inside the hospital, providing you with 5G connection and securing data transfer.

11.6.1 IoMT in the Hospital

The hospital's IoMT equipment could steadily evolve over time. In addition to basic vital signs, the band may track your oxygen levels, do routine electrocardiograms, and notify nurses if you've fallen. Your wristband may do quick tests and keep a watch on you like a nurse's aide. The hospital's IoMT gadgets may advance over time, much as how a smartwatch gets new features every year. The amount of time that workers may save might exponentially increase with a combination gadget like this wristband. It's a good idea to have a backup plan in case the backup fails. So, hospitals would greatly benefit from 5G's ability to reduce the cost of hospitals and other health-care labours while maintaining the same level of care. Significant cost reductions might be achieved by substituting non-invasive technologies for resource-intensive invasive methods [6–10].

11.6.2 IoMT at Home

The connection and speed of 5G might hasten the development of remote care monitoring (RCM) technology. And being able to quickly and reliably link medical information can be the difference between life and death. Take a patient with chronic renal illness as an example who might benefit from diligent blood monitoring to prevent kidney failure. A 5G-connected RCM device can send a warning to the patient and their doctor if a patient's blood pressure or glucose levels rise and emergency treatment is required. Prior to now, diabetics who require hourly glucose expert care could get these RCMs. The combined power of IoMT and 5G makes this scenario increasingly achievable, and it is expected to have a huge influence on our healthcare.

11.6.3 IoMT and 5G Innovations

The IoMT has previously been overused before to the arrival of COVID-19. Nonetheless, the coronavirus has increased interest in the ways that information may improve and restructure the health care system. Healthcare technology speculation has reached unprecedented heights this year. Now, the majority is fundamentally rethinking how we provide healthcare—not because it should, but because it must. When certain cities in our nation run out of clinic cots and there isn't always enough personnel to care for patients in need, we are happy to travel for replacements. Fortunately, designers and scientists are already anticipating how technology may advance healthcare, and 5G might provide the network needed to transform these concepts into reality [16–20].

Conclusion

5G networks will be essential for next-generation wearable health care devices. The 5G network is essential for healthcare and wearables using 5G technology from a functional and financial perspective. In this research, we highlighted a few 5G applications in wearable technology and healthcare. Furthermore carefully considered are various technological developments for achieving these standards in a 5G network. Academics now have additional possibilities for launching research projects in the area of 5G-based smart healthcare employing a variety of methodologies. Making wearables smaller and lighter is a big advantage for technology connected to health. More people are inclined to wear a stylish band than a big, bulky watch. Smart rings and "hearables," or smart earbuds, become increasingly alluring as they get smaller and lighter. Analytics might be used by healthcare professionals to assess data from a wearable device received via 5G for individuals with chronic health issues. The patient would get feedback right away on whether everything was alright, if an appointment was necessary, or, in the worst situation, whether they needed to call for emergency assistance. As a result, we have observed that the network "disappears" when used by big groups of individuals. These issues can be solved with 5G, which would help the wearable technology sector.

References

1. W. D. de Mattos and P. R. Gondim, "M-health solutions using 5gnetworks and m2m communications, "IT Professional, vol. 18, no. 3, pp. 24–29, 2016.
2. Chen AF, Kazarian GS, Jessop GW, *et al.* Robotic technology in orthopaedic surgery. J Bone Joint Surg Am. 2018; 100:1984–92.
3. Jin H, Hu Y, Tian W, *et al.* Kinematics and cooperative control of a robotic spinal surgery system. Robotica. 2016; 34:226–42.
4. Tian W, Wang H, Liu YJ. Robot-assisted anterior odontoid screw fixation: a case report. Orthop Surg. 2016; 8:400–4.
5. Zhang Q, Han XG, Xu YF, *et al.* Robot-assisted versus fluoroscopy-guided pedicle screw placement in transforaminal lumbar interbody fusion for lumbar degenerative disease. World Neurosurg. 2019;125: e429–34.
6. Pattichis CS, Kyriacou E, Voskarides S, *et al.* Wireless telemedicine systems: an overview. IEEE Antenna Propag Mag. 2002; 44:143–53.
7. Ullah H, Nair NG, Moore A, *et al.* 5G communication: an overview of vehicle-to-everything, drones, and healthcare use-cases. IEEE Access. 2019; 7:37251–68.
8. Latif S, Qadir J, Farooq S, Imran MA. How 5g wireless (and concomitant technologies) will revolutionize healthcare? Future Int. 2017; 9:93.
9. Börner Valdez L, Datta RR, Babic B, Müller DT, Bruns CJ, Fuchs HF. 5G mobile communication applications for surgery: An overview of the latest literature. Artif Intell Gastrointest Endosc 2021; 2(1).
10. Evans CR, Medina MG, Dwyer AM. Telemedicine and telerobotics: from science fiction to reality. Updates Surg. 2018; 70:357-362.
11. Nasser Alshammari *et al.*, "Technology-driven 5G enabled e-healthcare system during COVID-19 pandemic", ET Communications published by John Wiley & Sons Ltd on behalf of The Institution of Engineering and Technology, 2021;1–15.
12. Sigwele *et al.* "Energy Efficient Healthcare Monitoring System using 5G Task Offloading", Journal of Engineering and Scientific Research, Vol 1, Issue 2, December 2019.
13. Oleshchuk and Fensli, "Remote Patient Monitoring Within a Future 5G Infrastructure", Wireless Pers Commun, Springer Science+Business Media, LLC. 2010.
14. Singh *et al.* "Smart M-Health Continuous Monitoring System Using 5G Technology", Smart M-Health Continuous Monitoring System Using 5G Technology, December 2018.

15. A. Mishra and D. P. Agrawal, "Continuous health condition monitoringby 24×7 sensing and transmission of physiological data over 5-g cellular channels," in Computing, networking and communications (ICNC), 2015 international conference on. IEEE, 2015, pp. 584–590.
16. J. Lloret, L. Parra, M. Taha, and J. Tomas, "An architecture and protocolfor smart continuous ehealth monitoring using 5g," Computer Networks,vol. 129, pp. 340–351, 2017.
17. Abdul *et al.* "5G-based Smart Healthcare Network: Architecture, Taxonomy, Challenges and Future Research Directions", IEEE Access, Volume 4, 2016.
18. M.Lakha *et al.*, "Architecture of Real-Time Patient Health Monitoring Based on 5G Technologies", Advances in Science, Technology and Engineering Systems Journal Vol. 6, No. 2, 351-358 (2021).
19. Lalita Mishra *et al.* "Seamless Health Monitoring Using 5G NR for Internet of Medical Things", LLC, part of Springer Nature 2021.
20. Yangan Zhang *et al.* "Real-Time Remote Health Monitoring System Driven by 5G MEC-IoT", Electronics 2020, 9, 1753; doi:10.3390/electronics9111753
21. https://www.sciencedirect.com/science/article/abs/pii/S1389128617302189
22. https://www.frost.com/frost-perspectives/frost-sullivan-on-5g-takes-a-center-stage-with-huawei-at-mwc-2019-leading-the-way/

12

Cloud Assisted Data Storage in e-Healthcare System

A. Sathya[1]* and S. Kanaga Suba Raja[2]

[1]Department of Information Technology, Rajalakshmi Engineering College, Chennai, Tamil Nadu, India

[2]Easwari Engineering College, Chennai, Tamil Nadu, India

Abstract

The development of IPV6 and cloud computing has led to organizations and users employing the cloud for storing all kinds of information, including private data. Data privacy-related subjects are hotly contested because data is vital for research and has to be shared with sources that may be unknown or known. Therefore it is necessary to classify the sensitive attributes and securely store them in the cloud reducing the storage cost. The present encryption of cloud data is inefficient in terms of time consumption and cost. For data related to social media and health documentation, the major consideration is to protect the data's privacy. This is because, if the record owner's identity is unknown, the data would not be useful for any attacker. For efficient storage of all the attributes in the record, vertical partitioning is typically used wherein records are split vertically and connected by a common identifier such that none of the individually stored partitions can undo the privacy of the data. This work focused on a novel EPP-Attribute Classification. The method uses the concept of vertical partitioning to classify and store the sensitive attributes of the electronic health records across multiple locations increasing the storage performance of the cloud platform. Further, the classification comparison is done with the state-of-the-art algorithms to identify the best suitable classification algorithm for classifying the healthcare data. The results show that Naïve Bayes is the best suitable algorithm to classify the attributes, enhancing the data privacy-preservation approach in the cloud-assisted e-healthcare system.

Keywords: Cloud computing, privacy, sensitive attributes, encryption, healthcare

**Corresponding author*: arunachalam.sathyaa@gmail.com

A. Suresh, J. Ramkumar, M. Baskar and Ali Kashif Bashir (eds.) Resource Management in Advanced Wireless Networks, (247–266) © 2023 Scrivener Publishing LLC

12.1 Introduction

Due to the ubiquitous nature, cost-effectiveness, and robust nature of Cloud computing, data management in companies have been drastically altered. As sensitive data is sent to the cloud for data processing and storage, there are apprehensions about data privacy. Since users do not physically have the outsourced data, the accomplishment of data integrity and confidentiality is quite a challenging job. In the cloud, the data (which is usually encrypted) has to be stored and processed. As a result, the security specifications are not satisfactorily fulfilled by the cryptography-based data confidentiality and integrity protection techniques. When devising a secure resource sharing model utilizing a cloud platform, the privacy and security considerations of everybody involved should be ensured.

The process of analyzing fascinating, valuable, and unseen patterns from vast volumes of data stores and identifying the relationships among the patterns are defined as Data Mining [1]. From the data stores, data mining methods can draw out valuable data, which can also be, at times, sensitive. To protect this sensitive information, certain privacy-preserving methods can be employed on the original data before data mining. The goal of Privacy-Preserving Data Mining (PPDM) is to protect the sensitive information enclosed within the data. The data usability is not compromised when the PPDM conceals the individual sensitive information. K-anonymity, vertical partitioning is some of the popular techniques proposed to achieve PPDM. The distributed learning approaches can be utilized to operate on data that is vertically partitioned rather than analyzing data through centralization. In this situation, the data-processing algorithms are forwarded to each site and can only remit the analysis's outcomes instead of the original data.

K-anonymity is an attribute by which the released data is safeguarded from potential re-identification of the respondents to which the data refers. A fascinating feature of k-anonymity is its tie-up with protection methods that safeguard data integrity. When the data is vertically partitioned, different information about the same set of people or entities is collected by different organizations or sites. Entire attribute sets are required by these sites to collectively calculate a global data mining model and, at the same time, be unwilling to reveal data sets of their own.

Data storage security and data processing security are two characteristics that are related to privacy protection in cloud environments. The challenges of data processing security deal with protecting the user's privacy in a virtualized cloud platform during runtime. The challenges of data storage security are related to user data saved in data centers and in ensuring the privacy of these types of data. Extremely sensitive data is enclosed within healthcare information systems, and healthcare laws usually enforce the data privacy protection. Healthcare services use the cloud infrastructure to save and distribute medical documentation because of the rise in data amount and the cloud's beneficial attributes. Due to this, privacy-preserving techniques are employed to safeguard the data from undesired exposure. Individual traits and health conditions can be disclosed by genetic data. This type of data is personal, overly sensitive, and is also extremely recognizable. More than one individual is affected by genetic data, as blood relative's data can also be disclosed. Hence, it is particularly important to ensure the maximum usability of genetic data as well as avoid undesired exposure of the information [2].

Statistical analyses that integrate multiple distributed databases can be used as large applications in various situations. As an example, compared to individually analyzed databases, statistical models can be equipped to utilize more attributes or more records when the databases are integrated. Apprehensions related to confidentiality and data scale which comprises physical, regulatory, and legal obstacles, further impede the integration of data. Even if owners of databases work together to carry out integrated analyses, and nobody attempts to disclose the data confidentiality of others, these apprehensions are still there [3].

Databases that are horizontally partitioned garner a lot of interest in statistical literary works. These databases consist of similar numerical attributes for data subjects' disjoint sets. For instance, various state or local healthcare institutions may integrate the patients' data to enhance the accuracy of their general patient population's analyses. Analyses that are based on of sufficient statistics that are additive across the databases can be executed utilizing secure summation for the evaluation of these statistics. Not much consideration is given to vertically partitioned databases. These databases have similar data subjects but have diverse attribute sets. As an example, for the same organizations or individuals, a government institution may have the educational data, another institution may have the employment data. In contrast, the third institution may have health information.

Among data mining, clustering is a popular technique for analyzing health care records. The job of allocating a set of objects into clusters such that objects in the same cluster are more akin (in one way or the other) to each other when compared to objects in other clusters is defined as Cluster analysis or clustering. It is possible to discover general tendencies or trends from the clusters. Medical data analysis extensively utilizes cluster analysis. K-Means clustering algorithm is a Cluster analysis technique that attempts to divide n number of objects into k number of clusters. The clusters are made to optimize an objective partitioning benchmark, which is based on a similarity function like distance, such that, with regards to the database attributes, objects inside the same cluster are "similar", while objects of different clusters are "dissimilar". A distributed K-Means clustering protocol must retain each data site's privacy because of privacy regulations. It is quite a formidable task to retain each data site's privacy and, at the same time, execute K-Means clustering on distributed data. K-Means clustering algorithm employs an iterative refinement method. In K-Means clustering for distributed setting, if the individual data site responds to counting queries through output perturbation, output errors after round iterations get collected, and this, in turn, may influence the K-Means clustering's precision [4]. Another popular technique used in Data Mining is the classification technique, which is a supervised learning method wherein a set of independent attributes are mapped to an output attribute called the class label. The principle of K Nearest Neighbor instead of the K-Means algorithm can be effectively used for classification. The primary objective of the proposed EPP (Euclidean L3P-based Multi-Objective Successive Approximation Privacy Preservation) Attribute Classification is the hybridized classification of genetic algorithm and naïve bayes algorithm. Here vertical partitioning was used to classify and store the sensitive attributes of the electronic health records across multiple locations increasing the storage performance of the cloud platform.

12.1.1 Our Contribution

This work investigated techniques to improve privacy-preserving in the storage of electronic health records in the cloud-assisted e-Healthcare system by performing

- A new vertical partitioning technique using partition wise anonymization and perform attribute classification
- Sectional encryption techniques to secure data.
- Ensuring the privacy of the patient's data in the heterogeneous cloud.
- Evaluate the performance with metrics such as Precision, Recall, F-measure, and Accuracy.
- Comparing and identifying the suitable classification algorithm for the healthcare database.

The associated literary works have been discussed in Section 12.2. The proposed techniques have been described in Section 12.3. The experimental results have been studied in Section 12.4. Finally, the conclusion of the work has been explained in Section 12.5.

12.2 Related Works

In the study of epidemiology, the practicability of privacy-preserving data mining was analyzed by Kikuchi *et al.* [6]. For the data mining algorithm, linear multiple regression is utilized to detect the major criteria within numerous potential variables like the history of several diseases. The linear model is used to quantify the most important reason for death from the distributed dataset, which is in terms of the disease and patient data. An experiment was performed on a real medical dataset related to a stroke, and an effort was made for the application of multiple regression with six predictors of sex, age, the medical scales, like, the modified Rankin Scale and Japan Coma Scale. The work consisted of the following inputs: (1) a feasible privacy-preserving protocol was proposed for linear multiple regression with datasets which are vertically partitioned, (2) the proposed system's practicability was demonstrated utilizing the real medical dataset distributed into two parties, which are, the local government agency who is aware of the patient even after they have been discharged from the hospital and the hospital who is aware of the disease's technical details while the patients are admitted there, (3) the PPDM system's performance and precision were demonstrated in order to assess the anticipated time for processing when an arbitrary number of predictors are utilized and (4) the complexity of the extended models of vertical partition was also analyzed.

Sun *et al.* [7] suggested an ingenious infrastructure that aided the safe and privacy-preserving analysis of personal health information from various providers with different governance policies. The purpose of the infrastructure is to examine the relationship between healthcare expenses and Type 2 Diabetes Mellitus status. The proposed technique comprises the application of distributed machine learning to study information that is vertically partitioned from a potential population-based cohort study knows as the Maastricht Study and data from Statistics Netherlands. An optimal solution that considered the legal/ethical, technical, and scientific issues was explored. This work dealt with detailing these issues, how these issues are addressed in a practical application case, and an experimental simulation was also performed.

A protocol to perform secure regressions and similar analyses on data, which is vertically partitioned were submitted by Karr *et al.* [3]. Data owners can use this protocol to evaluate a linear regression's standard errors and coefficients and for the analysis of regression model diagnostics, without revealing their attribute values with one another. There is no involvement of third parties. This protocol is utilized to execute other processes for which the sufficient statistics are the sample's covariance and means. This protocol is based on a secure matrix multiplication algorithm, which is utilized by pairs of owners to evaluate the full data covariance matrix's off-diagonal blocks.

A privacy-preserving distributed data mining scheme was suggested by Vaidya & Clifton [8]: a global model can be developed by the collaboration of data sources without revealing their information to others. Naïve Bayes is generally utilized as a baseline classifier to constantly give a classification performance, which is feasible. The work brought privacy-preservation to Naïve Bayes classification on data, which is vertically partitioned.

Frequent itemset mining and association rule mining are two extensively analyzed and well-known techniques of data analysis for a wide application range. Privacy-preserving mining on databases which are vertically partitioned was studied by Geetha *et al.* [9]. In these scenarios, the data owners want to study the frequent itemsets or association rules from a collective dataset with the least amount of disclosure of their raw and data to third parties and other data owners. A secure comparison scheme and an effective homomorphic encryption scheme was designed in this work. A cloud-aided frequent itemset mining solution was also proposed to be utilized to construct an association rule mining solution. The solutions in this work were developed for outsourced databases to let various data owners to effectively share their data in a secure manner compromising the privacy of the data. Compared to most of the existing solutions, these solutions also disclose minimal information related to the raw data. The performance of the proposed solutions is 3 to 5 orders of magnitude higher when compared to the only known solution which obtained the same level of privacy as these proposed solutions. From the experimental results based on utilizing different datasets and parameters, the work is able to show that each of these solutions' run time is only one order higher compared to the run time of the best non-privacy-preserving data mining algorithms. The resource usage at the data owner's end is quite low as both the computing work and data are outsourced to the cloud servers.

The scheme to construct a hierarchical categorical clustering algorithm on the vertical and horizontal partitioned dataset was provided by Sheikhalishahi & Martinelli [10]. When the data is distributed between two parties, it is presumed that for the sake of general advantages, both parties want to identify the whole dataset's clusters, however, due to apprehension about privacy, the original datasets are not shared. This work proposed algorithms which are based on secure number comparison protocol and secure weighted average protocol, to safely evaluate the desired criteria for the development of the scheme of the cluster.

For the distributed greedy algorithm, Jung *et al.* [11] recommended a structure where the examples in the feasible set are obtained from various parties. In this structure, the majority of the generic distributed greedy algorithms have been transformed into a privacy-preserving type that is able to get the same outcome as the original greedy algorithm. However, private data related to the examples is still safeguarded.

In numerous different fields, an important problem in data mining is the classification. Yu *et al.* [12] try to construct a design to forecast an attribute (binary). A secure and effective privacy-preserving algorithm is to aid in vector machine (SVM) classification over vertically partitioned data is recommended. Effective homomorphic encryption and a secure identification design were formulated by Li *et al.* [13]. Later, a cloud-aided frequent itemset mining solution is suggested to generate an association rule mining solution. Compared to the existing solutions, the proposed solution discloses more limited data in terms of the raw data. The proposed solution's performance is 3 to 5 orders of magnitude higher. Compared to the best non-privacy-preserving data mining algorithms, the runtime of each of these solutions' is proved to be only one order higher. Since the data, as well as the computing action, as well as the computing action, is outsourced to the cloud servers, at the data owner's end, the resource usage is deficient.

Vaidya & Clifton [14] put forward a K-Means clustering system when different sites consist of different attributes for a standard set of entities. Each site studies the private entity's cluster to establish the features at diverse sites. Vaidya & Clifton [15] came up with the ID3 algorithm's generalized privacy-preserving variant for vertically partitioned data, which is split over two or more parties. Extensive confirmation of security is provided, which permits a strong bond on the disclosed data. Jagannathan & Wright [16] submits two additives to privacy-preserving data mining. For the concept of arbitrarily partitioned data, there is a generalization of data which are vertically and horizontally partitioned. For arbitrarily partitioned data, an effective privacy-preserving protocol is executed for K-Means clustering. Dwork & Nissim [17] study two basic directions: vertically partitioned databases and multi-attribute databases, in which different databases have the accumulation of different attribute subsets.

Sheela & Vijayalakshmi [18] proposes a disturbance and anonymization procedure worked on the vertically partitioned data. A third-party coordinator is employed to partition the data recursively in different parties. The parties disturb the data by determining the mean when the designated threshold level is attained. The disorder controls the statistical correlation with attributes.

12.3 Research Methodology

This section details the preprocessing phase of the proposed EPP Attribute Classification (EPP-AC) technique using the sectional encryption to classifying the common and sensitive attributes and identifying which classification algorithm suits better for the proposed cloud-assisted e-healthcare system. The framework of the proposed technique is shown in Figure 12.1. Once the electronic health records are split vertically, data generalization was then applied on the vertically split data and the vertically split data with the highest vulnerability was encrypted using the Data Encryption standard to improve stand alone privacy. The proposed is the hybridization of the genetic algorithm.

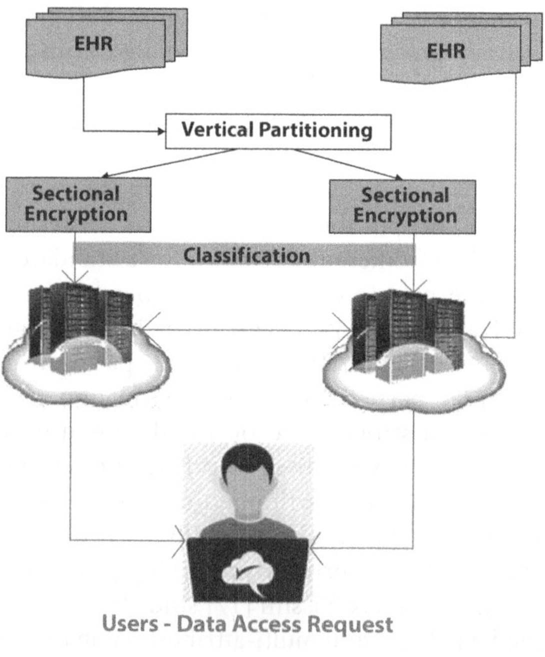

Figure 12.1 Framework of the proposed EPP – attribute classification.

12.3.1 EPP Attribute Classification and Storage

The electronic health record searching process shows a concurrent examination of the solutions space, providing the iterative updating of the probability array, which stores information concerning the bests solutions found so far. Moreover, the records consists of both the sensitive and common attributes. Common attributes can be publicly viewed whereas sensitive attributes are encrypted and stored as a privacy preserved database. And the access of it has to satisfy the specified access control requirements. For instance, consider the databases of four hospitals each having 14 sensitive attributes. Though the number of attributes remains the same the total number of records gets varied for each healthcare center and this is completely dependent on the number of patients registered for a particular healthcare center to undergo the medical treatment. Here the concept of vertical partitioning was used to classify and store the sensitive attributes of the electronic health records across multiple locations increasing the storage performance of the cloud platform. An unique id was created for each record of sensitive attributes and this was used as the key to connecting two records split vertically to obtain the original record. Let us suppose an optimization problem in which a feasible solution may be described as a binary string. So the probability array might then take the form of a vector. Each entry stores a probability connected with a position with each bit of the optimal solution. If a given probability entry addresses or converges to one that would be the associated value of the corresponding bit in the optimal solution. Each probability (Pi) of the vector represents a single binary attribute of the solution. At the beginning of the searching process, the whole vector of probabilities is set to 1/2, providing a uniform and comprehensive exploration of the solutions space, however, this can lead to the local optima problem. In the proposed work a greedy approach is used to find optimal solutions for a random set of solutions and these become the premises for the initial set of records. Once the initial population is identified a weight is assigned for each solution based on the fitness obtained. This process improves the searchability and poorer solutions move towards the cumulative better solution. For each algorithm iteration, a new population of potential solutions is performed by the probabilities in the vector (i.e., those probabilities close to zero, are prone to create solutions with the corresponding bit set to zero and vice versa).

This work proposed EPP-AC is the pre-processing stage of the technique EPP technique for cloud-based privacy-preserving mechanisms in the ehealthcare system. The proposed technique uses the vertical split mechanism to divide the dataset into multiple subsets and ensure that the privacy constraints are met. Once the constraint conditions are met, it is proposed to optimize the split and identify the data subset that has to be encrypted. In EPP-AC, if the population of the p individuals found in the search space is represented as a matrix as in Equation (12.1):

$$X = \begin{pmatrix} x_{11} & x_{12} & \dots & x_{1l} \\ x_{21} & x_{22} & \dots & x_{2l} \\ . & . & . & . \\ . & . & . & . \\ x_{\mu 1} & x_{\mu 2} & \dots & x_{\mu l} \end{pmatrix} \tag{12.1}$$

where, the i-th individual is represented by row $x_{i\bullet} = (x_{i1}, x_{i2}, \dots, x_{il})$ (i = 1,….,μ), and the gene pool by column $x_{\bullet j} = (x_{1j}, x_{2j}, \dots, x_{\mu j})'$ (j = 1,….,l). During the iteration, the offspring/child (y) is obtained through random genes from the gene pool after the greedy search. An weighted probability is assigned to all the gene pools. Thus, the attributes of each gene pool is assessed, and the offspring is produced by extracting a component y_j belonging to the binomial distribution together with $P\{y_j = 1\} = p_j$ and $P\{y_j = 0\} = 1 - p_j$ *for* $j = 1,\dots,l$. Thus, only the vector $p \in (0, 1)^l \subset R^l$ is required to specify the population, and the X offspring is produced using $\mu < \lambda$. A p best offspring is chosen from the gene pool to provide chances to discover the new vector of attributes.

The following updating rule will be considered by the CBCL which is similar to PBIL: If $y_{1:\lambda}, y_{2:\lambda}, \dots, y_{\lambda:\lambda}$ indicates λ, the sorted trial vectors together with $f(y_{1:\lambda}) \le f(y_{2:\lambda}) \le \dots \le f(y_{\lambda:\lambda})$, where, the minimized objective function is f. The $\alpha \in (0,1)$ and $t \in N_0$ update rules are as per Equation (12.2) and Equation (12.3):

$$p^{(t+1)} = \frac{1}{\mu}\sum_{k=1}^{\mu} y_{k:\lambda}^{(t)} \tag{12.2}$$

$$p^{(t+1)} = (1-\alpha)p^{(t)} + \alpha\frac{1}{\mu}\sum_{k=1}^{\mu} y_{k:\lambda}^{(t)} \tag{12.3}$$

Where, $p_i^{(0)} = w_i \ for \ i = 1,...,l$ represents the initial setting. The evolutionary algorithm, which exploits the gene pool recombination with the selection and the EPP-AC is used for association. The updated rules will result in a Markov chain-modelled stochastic algorithm.

Algorithm-EPP Attribute Classification

1. *Obtain Electronic Health Records*
2. *Identify record which are quasi-identifiers*
3. *List Quasi Identifiers which can narrow down the patient identity*
4. *Randomly split vertically into sensitive datasets linked by unique id*
5. *Generalize data in each dataset on the quasi identifiers.*
6. *Identify vertical split group with highest number of quasi-identifier*
7. *Encrypt using DES*

a. *Plain text → DES Algorithm → 64 bit cyphertext*

2. *Split the data using ten fold cross validation*
3. *Train Naïve Bayes Classifier using training data*
4. *Test classifier performance with test data*
5. *Compare the obtained results*
6. *Classify generalized data and measure benchmark parameters*

Quasi identifier

Quasi Identifiers are a set of attributes selected with an external composition which can be used to re-identify all/some of the entities linked to the referred information. When the quasi identifier are used together, it can identify an individual record. The quasi-identifiers used with respect to health information are age, gender, zip codes, regions, race, and ethnicity.

Dataset - Vertical Partitioning

In vertically partitioned data, a particular individual's various attributes are distributed over a set of data sources. In data that is horizontally partitioned, the outcomes of the data analyses are later integrated. However, in vertically partitioned data, all data providers are unable to perform the whole analysis independently of the others. This is quite difficult either when there is an unavailability of the unique identifier or when there is a legal restriction to use a unique identifier to link records across multiple data providers. To efficiently focus on this difficulty, a high level of technical finesse is needed to address the privacy and/or legal constraints together. In this work, Vector V of random numbers from a uniform random distribution is created. Vertical partitioning of a table T splits it into three tables referred to as sub-tables, each of which contains a subset of the columns in T. The distributed learning approaches can be utilized to operate on data that is vertically partitioned rather than analyzing data through centralization. In this situation, the data-processing algorithms are forwarded to each site, and can only remit the analysis's outcomes instead of the original data.

Generalization

In generalization, the specific value in the data is replaced by a general value. Though the generalization of data lowers the data quality, it still retains some of its semantic information. The generalization can be seen as a non-deterministic polynomial problem.

Sectional Encryption

The DES initially started with a 64-bit key. Later, the NSA restricted the DES to a 56-bit key length. As a result, DES removes 8 bits from the 64-bit key and later utilizes the compressed 56-bit key, which was obtained from the 64-bits key to encrypt the data in a block size of 64-bits. DES is flexible as it can work in various modes like OFB, CFB, ECG, and CBC. The Electronic Frontier Foundation (EFF) and distributed.net work together in January 1999 to openly break within 22 hours and 15 minutes the DES key. Even though there are hypothetical attacks, the algorithm is considered to be almost secure in the Triple-DES form. The Advanced Encryption Standard (AES) has taken over from the cipher in recent times [19, 20].

Naïve Bayes Classification

The de facto classifier utilized for classifying text is the Naïve Bayes classifier, which is an effective and uncomplicated baseline classifier. This is based on the Bayesian formulation of the classification problem that utilizes the simplifying assumption of attribute independence. This is enforced easily and utilized to provide excellent outcomes. As a result, when the initial evaluation is performed utilizing the Naïve Bayes classifier to function as a baseline and also determine if the solutions need to be elaborated further [24]. The Naïve Bayes algorithm has exhibited excellent performance and is also extensively utilized to classify documents. It can handle both continuous and discrete data. The fundamental concept of the Naïve Bayes algorithm is to utilize the joint probabilities of categories and words to assess the probabilities of categories provided a document. This algorithm evaluates the posterior probability that the document is a part of different classes and allocates it to the class with the maximum posterior probability. The Bayes rule is utilized to evaluate a class's posterior probability, and a class with the maximum posterior probability is allocated for the testing sample. For the Naïve Bayes algorithm, the naive part is the assumption of word independence that within a category, the given the word's conditional probability is considered to be autonomous from the conditional probabilities of other words provided in that category.

12.4 Results and Discussion

12.4.1 Data Collection

This section presents the simulation results of the proposed EPP-attribute classification for privacy preservation in the cloud-assisted eHealth environment. The proposed research study evaluated using a standard heart failure database from the Kaggle repository having 3 million instances in total. The Heart Disease dataset used created from the University of California in Irvine, California (UCI) Center for Machine Learning and Intelligent Systems. It consists of four databases from four hospitals. Each database has 14 features but the number of records is different. The concept of vertical partitioning was used to classify and store the sensitive attributes of the electronic health records across multiple locations increasing the storage performance of the cloud platform. An unique id was created for each record of sensitive attributes and this was used as the key to connecting two records split vertically to obtain the original record.

12.4.2 Experimental Setup

Experiment was conducted using a Xeon E5-2640 Server with 128Gb DDR4 memory running on VMware and Linux OS with 6 Virtual machines were used for the experiments. Simulations were carried out for 15 runs. Table 12.1 shows the simulation parameters and Figure 12.2 to 12.5 shows the results of classification accuracy, precision, recall, and F Measure respectively.

Table 12.1 Simulation parameters.

Partitioning type	Vertical
Encryption type	Symmetric
Encryption method	DES
Number of VM	6
VM type	Heterogeneous

12.4.3 Proposed Results

From the Figure 12.2, it can be observed that the naive bayes-proposed vertical partitioning has higher classification accuracy by 1.63% for K Means, by 0.33% for naive bayes, by 5.53% for K Means - vertical partitioning, by 2.1% for naive bayes - vertical partitioning, by 2.71% for K Means - PSO vertical partitioning, by 0.57% for naive bayes - PSO vertical partitioning and by 2.15% for K Means - proposed vertical partitioning respectively.

The proposed technique matches the precision of the original data for different classifiers. It also shows the improvement in the precision

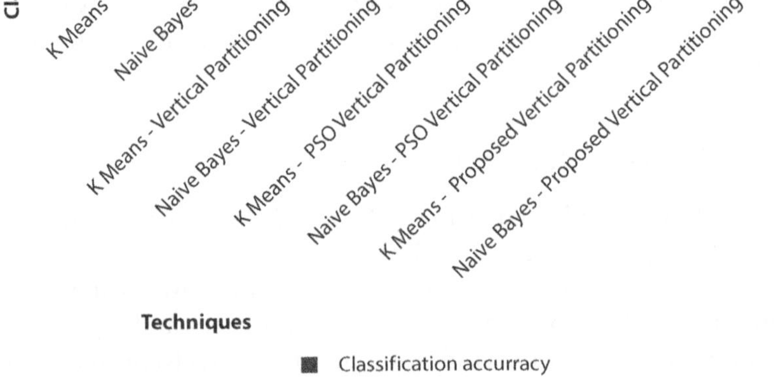

Figure 12.2 Classification accuracy for proposed techniques compared with original data.

compared to regular vertical partitioning. Similarly. There is no significant drop in the recall values compared to the raw data. From the Figure 12.3, it can be observed that the naive bayes-proposed vertical partitioning has higher average recall by 0.69% for K Means, by 0.12% for naive bayes, by 4.33% for K Means - vertical partitioning, by 1.34% for naive bayes - vertical

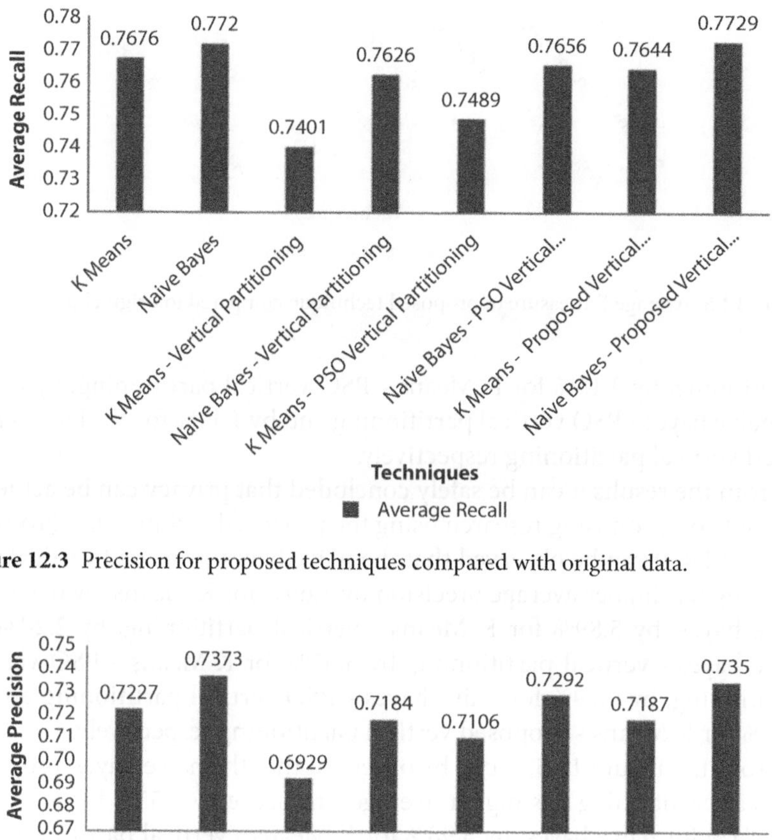

Figure 12.3 Precision for proposed techniques compared with original data.

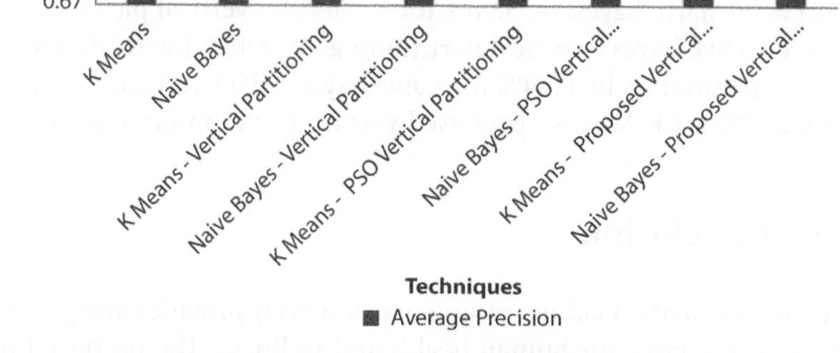

Figure 12.4 Average recall of proposed technique compared to original data.

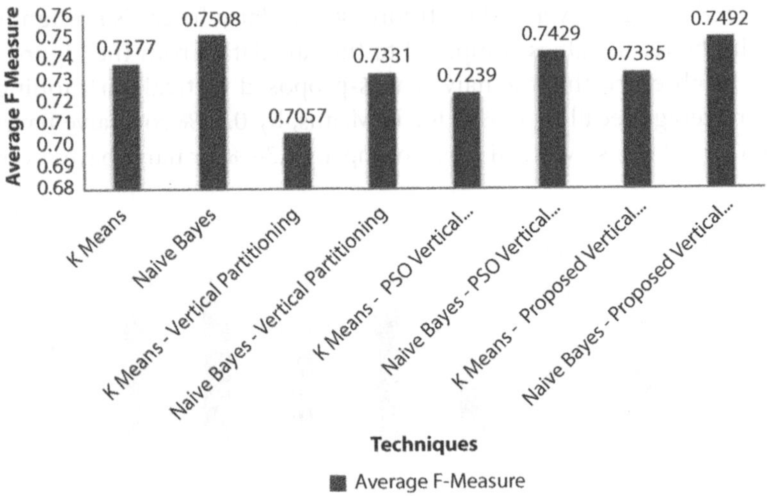

Figure 12.5 Average F-measure of proposed technique compared to original data.

partitioning, by 3.15% for K Means - PSO vertical partitioning, by 0.95% for naive bayes - PSO vertical partitioning and by 1.11% for K Means - proposed vertical partitioning respectively.

From the results it can be safely concluded that privacy can be achieved without compromising research using the proposed architecture. From the Figure 12.4, it can be observed that the naive bayes-proposed vertical partitioning has higher average precision by 1.69% for K Means, by 0.31% for naive bayes, by 5.89% for K Means - vertical partitioning, by 2.28% for naive bayes - vertical partitioning, by 3.37% for K Means - PSO vertical partitioning, by 0.79% for naive bayes - PSO vertical partitioning and by 2.24% for K Means - proposed vertical partitioning respectively.

From the Figure 12.5, it can be observed that the naive bayes-proposed vertical partitioning has higher average f measure by 1.55% for K Means, by 0.21% for naive bayes, by 5.98% for K Means - vertical partitioning, by 2.17% for naive bayes - vertical partitioning, by 3.43% for K Means - PSO vertical partitioning, by 0.84% for naive bayes - PSO vertical partitioning and by 2.12% for K Means - proposed vertical partitioning respectively.

12.5 Conclusion

The cloud-assisted e-healthcare system extensively provide timely and reliable medical service for human health and wellness. The proposed novel framework is the pre-processing stage of the EPP access control technique utilizing the strength of encryption and generalization to vertically partition the datasets and classifying the common and sensitive attributes for maintaining the privacy preservation before its storage in the cloud platform. This work used a greedy approach for data generalization but further work need to be carried out using meta-heuristic algorithm to overcome the local optima problem faced by greedy search algorithms. Naïve Bayes classifiers have fast decision making and do not necessitate large dataset for learning. The result shows that naive bayes-proposed vertical partitioning has higher classification accuracy by 1.63% for K Means, by 0.33% for naive bayes, by 5.53% for K Means - vertical partitioning, by 2.1% for naive bayes - vertical partitioning, by 2.71% for K Means - PSO vertical partitioning, by 0.57% for naive bayes - PSO vertical partitioning and by 2.15% for K Means - proposed vertical partitioning respectively. Thus proving that naïve bayes is the best suitable algorithm to classify the common and sensitive attributes for the data privacy-preservation in the cloud-assisted e-healthcare system. Also, vertical partitioning has significantly enhanced the EPP technique by classifying and storing the sensitive attributes of the electronic health records across multiple locations increasing the storage performance and privacy measure of the cloud platform.

References

1. Kaur, R., Sharma, M., & Taruna, S. (2019, March). Privacy Preserving Data Mining Model for the Social Networking. In *2019 6th International Conference on Computing for Sustainable Global Development (INDIACom)* (pp. 763-768). IEEE.
2. Mendes, R., & Vilela, J. P. (2017). Privacy-preserving data mining: methods, metrics, and applications. *IEEE Access, 5*, 10562-10582.
3. Karr, A. F., Lin, X., Sanil, A. P., & Reiter, J. P. (2009). Privacy-preserving analysis of vertically partitioned data using secure matrix products. *Journal of Official Statistics, 25*(1), 125.
4. Yi, X., & Zhang, Y. (2013). Equally contributory privacy-preserving k-means clustering over vertically partitioned data. *Information systems, 38*(1), 97-107.
5. Baluja, S. (1994). Population-based incremental learning. a method for integrating genetic search based function optimization and competitive learning. Carnegie-Mellon Univ Pittsburgh Pa Dept Of Computer Science.
6. Kikuchi, H., Hamanaga, C., Yasunaga, H., Matsui, H., Hashimoto, H., & Fan, C. I. (2018). Privacy-preserving multiple linear regression of vertically partitioned real medical datasets. *Journal of Information Processing, 26*, 638-647.

7. Sun, C., Ippel, L., van Soest, J., Wouters, B., Malic, A., Adekunle, O., ... & van Oppen, C. (2019). A Privacy-Preserving Infrastructure for Analyzing Personal Health Data in a Vertically Partitioned Scenario. *Studies in health technology and informatics, 264*, 373-377.
8. Vaidya, J., & Clifton, C. (2004, April). Privacy preserving naive bayes classifier for vertically partitioned data. In *Proceedings of the 2004 SIAM International Conference on Data Mining* (pp. 522-526). Society for Industrial and Applied Mathematics.
9. Geetha, K., Guptha, K. G., & Prasad, S. N. V. A. S. R. K. (2017). Privacy-Preserving Outsourced Association Rule Mining on Vertically Partitioned Databases.
10. Sheikhalishahi, M., & Martinelli, F. (2017, July). Privacy preserving clustering over horizontal and vertical partitioned data. In *2017 IEEE Symposium on Computers and Communications (ISCC)* (pp. 1237-1244). IEEE.
11. Jung, T., Li, X. Y., & Zhang, L. (2013). A general framework for privacy-preserving distributed greedy algorithm. *arXiv preprint arXiv:1307.2294.*
12. Yu, H., Vaidya, J., & Jiang, X. (2006, April). Privacy-preserving svm classification on vertically partitioned data. In *Pacific-Asia Conference on Knowledge Discovery and Data Mining* (pp. 647-656). Springer, Berlin, Heidelberg.
13. Li, L., Lu, R., Choo, K. K. R., Datta, A., & Shao, J. (2016). Privacy-preserving-outsourced association rule mining on vertically partitioned databases. *IEEE Transactions on Information Forensics and Security, 11*(8), 1847-1861.
14. Vaidya, J., & Clifton, C. (2003, August). Privacy-preserving k-means clustering over vertically partitioned data. In *Proceedings of the ninth ACM SIGKDD international conference on Knowledge discovery and data mining* (pp. 206-215).
15. Vaidya, J., & Clifton, C. (2005, August). Privacy-preserving decision trees over vertically partitioned data. In *IFIP Annual Conference on Data and Applications Security and Privacy* (pp. 139-152). Springer, Berlin, Heidelberg.
16. Jagannathan, G., & Wright, R. N. (2005, August). Privacy-preserving distributed k-means clustering over arbitrarily partitioned data. In *Proceedings of the eleventh ACM SIGKDD international conference on Knowledge discovery in data mining* (pp. 593-599).
17. Dwork, C., & Nissim, K. (2004, August). Privacy-preserving datamining on vertically partitioned databases. In *Annual International Cryptology Conference* (pp. 528-544). Springer, Berlin, Heidelberg.
18. Sheela, M. A., & Vijayalakshmi, K. (2017). Partition based perturbation for privacy preserving distributed data mining. *Cybernetics and Information Technologies, 17*(2), 44-55.
19. Karthik S, Muruganandam A. Data encryption and decryption by using triple DES and performance analysis of crypto system. International Journal of Scientific Engineering and Research. 2014;2(11):24-31.
20. Wahid, M. N. A., Ali, A., Esparham, B., & Marwan, M. (2018). A Comparison of Cryptographic Algorithms: DES, 3DES, AES, RSA and Blowfish for

Guessing Attacks Prevention. *Journal Computer Science Applications and Information Technology, 3*, 1-7.
21. Mandala, J., & Rao, M. C. S. (2019). Privacy preservation of data using crow search with adaptive awareness probability. Journal of information security and applications, 44, 157-169.
22. Asta, S., & Uyar, A. Ş. (2012). A Novel Particle Swarm Optimization Algorithm. Computer & Informatics Faculty Istanbul Technical University, Istanbul, Turkey.
23. X. Meng, J. Li, M. Zhou, X. Dai, and J. Dou, "Population-based incremental learning algorithm for a serial colored traveling salesman problem," IEEE Transactions on Systems, Man, and Cybernetics: Systems, vol. 48, pp. 277–288, Feb 2018.
24. Malik., A, Sharma., A, Vinod., S. (2013) "Greedy Algorithm", International Journal of Scientific and Research Publications, Volume 3, Issue 8, August 2013.
25. Khodaparast, F., Sheikhalishahi, M., Haghighi, H., & Martinelli, F. (2019, October). Privacy-Preserving LDA Classification over Horizontally Distributed Data. In *International Symposium on Intelligent and Distributed Computing* (pp. 65-74). Springer, Cham.
26. Said, A. (2005, September). Measuring the strength of partial encryption schemes. In IEEE International Conference on Image Processing 2005 (Vol. 2, pp. II-1126). IEEE.
27. Cheng, H., & Li, X. (2000). Partial encryption of compressed images and videos. IEEE Transactions on signal processing, 48(8), 2439-2451.

13

Challenges in 5G Building Sustainable Networks and Its Opportunities

Sharanya S.

Department of Data Science and Business Systems, School of Computing, SRM Institute of Science and Technology, Kattankulathur, India

Abstract

The rampant deployment of 5G has already begun in developed countries. There have been severe acquisitions from the environmentalists about the negative impacts of 5G on the earth's flora and fauna. This chapter presents the elemental 5G technologies and analyses its impact on the ecosystem. In addition to this, the work attempts to demystify the myths about the 5G and discusses the scope for 5G to transform into a more sustainable infrastructure. The chapter also contributes the Green 5G solutions and policies that are conceived and implemented by various network providers and researchers to build green network. The latter part of this work highlights the potential areas that could be explored to build green and clean 5G with their design challenges.

Keywords: Small cells, massive MIMO, beam forming, millimeter waves, green network

13.1 Introduction to 5G Technologies

5G technology is perceived as an important milestone under the umbrella of digital single market strategy for dense connectivity of heterogeneous devices. The rationale behind the launch of 5G is to achieve more powerful "gigabit society" by 2025. As the world is envisioning rampant increase in the number of devices connecting to the internet, there is a high demand for networks with high prowess. The European Commission has highlighted

Email: nice_sharanya@yahoo.co.in

A. Suresh, J. Ramkumar, M. Baskar and Ali Kashif Bashir (eds.) Resource Management in Advanced Wireless Networks, (267–290) © 2023 Scrivener Publishing LLC

the below mentioned connectivity targets to be realized through 5G technologies [1]:

i. To provide high speed internet access in the range of 1GBPS to all public service providers
ii. To provide download speed of atleast100 MBPS for urban and rural households
iii. To provision 5G coverage along roadways and railways

5G is culmination of bunch of new technologies in the existing 4G networks. The increase in demand for high rate data movement, low latency, denser connections, and devices capable of connecting to internet are the major driving factors for conceiving the notion of 5G. Some of the projected application areas of 5G is illustrated in Figure 13.1.

5G technologies has not emerged completely. The development of 5G is attributed to atleast five new technologies that uses sophisticated hardware to attain the performance goals of 5G [2]. Energy efficiency is an important design consideration in planning and decommissioning 5G networks. The trend line shows the provisioning of newer technologies and strategies

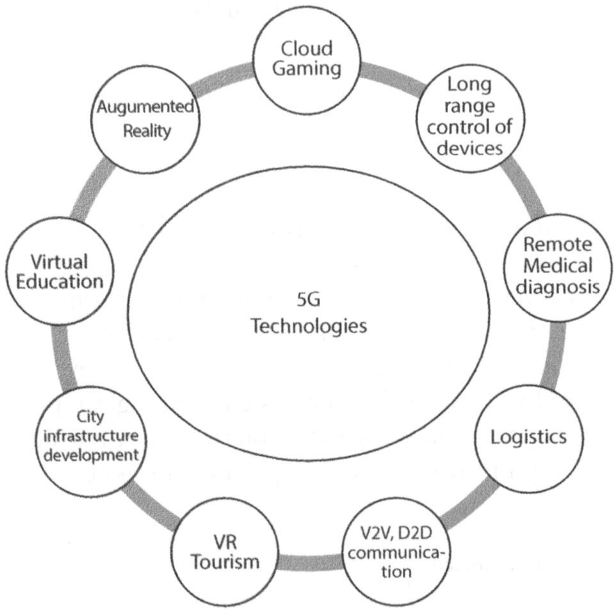

Figure 13.1 Some of the application areas of 5G technologies.

starting from smart power to predictive maintenance, which will evolve into a more sustainable network [3].

The evolution of 5G is characterized by five featuring technologies namely millimeter waves, massive Multiple Input Multiple Output (MIMO), small cells, beam forming and full duplex communications [4].

i) Millimeter waves (mm)

The demand for the high data rate with assured network QoS has led to the proposal of short range waves with frequencies higher than 6GHz. These waves fall under the millimeter wave category which ranges between [30-300 GHz] [4]. These waves can travel only in Line of Sight (LoS) of the transmitter to a maximum coverage distance of 10.8 km. The coverage range may shrink further, when the LoS is blocked by buildings or other objects. Figure 13.2 shows the location of mm waves in the spectrum.

Transmission of these waves demand packaging multiple antenna elements that operates at millimeter range, thus resulting in formation of narrow beams [5]. Though these waves promises wider bandwidth and formation of dense networks, they are more likely to suffer from two major bottlenecks namely LOS requirement and limited range. Apart from these, the millimeter waves are rainfading as they get absorbed by moisture in the atmosphere. These limitations are overcome by deploying high gain antenna elements that can enhance the net radiation power [6].

ii) mMIMO

Massive MIMO (mMIMO) uses array of antenna elements (more likely to be 32) for transmitting and receiving data and is shown in Figure 13.3. The Base Station (BS) deploys non repetitive precoding for each data stream. The location of the User Equipment (UE) is used to determine the signal optimization for the particular UE with reduced interference. As the 5G BS adopts nMIMO, multiple datastreams are formed

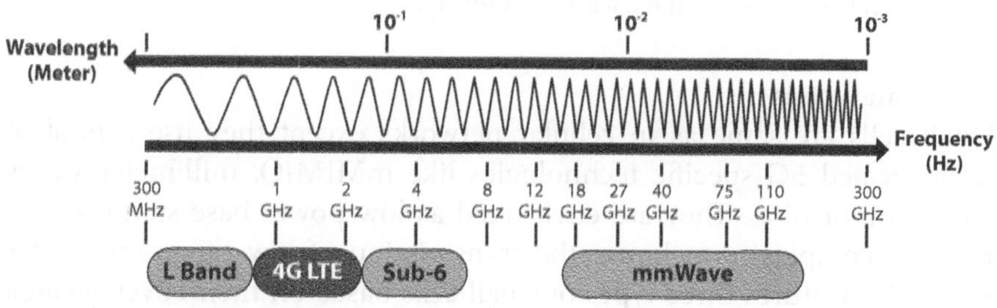

Figure 13.2 mm Waves in the spectrum band [7].

Figure 13.3 mMIMO.

such that there is no interference between each other by following unique pattern of radiation [8].

Though the mMIMO is a captivating technology, installation of multiple transceivers requires additional hardware cost and increases the complexity in the base station. A major challenge in deploying mMIMO and precise beamforming is the requirement of complex real time planning to scale up the array of antenna elements, UE and bandwidth considerations. This in turn give rise to the demand for sophisticated load balancing algorithms that utilizes the hardware efficiently.

iii) Small cells

Small cells are same as in cellular networks except they use a pool of sophisticated 5G specific technologies like mMIMIO, millimeter waves and beamforming. They are envisioned as low power base stations with reduced complexity to house the transmission of low range millimeter waves. 5G features three types of small cells based on their coverage area namely femtocells (10 m), picocells (200 m) and microcells (around 2 km). The transceivers or BS in small cells are very compact that they can be mounted on walls or even on lamp posts. The backhaul connections (connections between the BS and core network) can be established through heterogeneous mode such as fiber optical cables, microwave or wired connections [9].

The major limitations of small cells are short coverage area, hence needs more number of transceivers has to be deployed. In addition to this, the cost of installation of small cells is expensive than 4G cells.

iv) Beamforming

This is a subset of mMIMO that uses multiple antennas to form beams. The direction of transmission controlled through beam forming by adjusting the antennas based on the phase and signal magnitude. Same signal is transmitted multiple times with phase shift between them. Multiple copies of same data will be received by the receiver [10]. This process boosts the signal strength thus supporting elevated number of customers. The signals can be directed in desired direction, hence the interference to unintended receiver BS can be avoided.

Though there are numerous advantages of deploying beamforming in 5G, the complete setup is highly complex. Also, the study about energy requirements of beamforming in 5G is still in nascent stage.

v) Full Duplex Communications

Full duplex communication at a transceiver is achieved by exploiting the reciprocity property at the physical layer. The radio waves travel back and forth in same frequency for transmitting and receiving the waves simultaneously. This roughly doubles the speed of wireless communications to a great extent [11].

The major hurdle faced while implementation of full duplex communication is the pesky echo, which is caused due to the interference between the transmitting and receiving radio waves. Canceling the interference due to self-transmitted signals is done at multiple levels namely at antennas, circuits and digital baseband.

Thus 5G technologies has evolved from the previous generations by adding new prowess. However, there is a constant note from the environmentalists and biologists about the impact of 5G deployment. This article focuses on scientifically demystifying the myths on 5Gby throwing light areas that the 5G technology vendors should focus to make 5G a greener and cleaner technology.

The organization of the article is as follows: Section 13.2 reviews the literature on hazardous impact of 5G on the flora and fauna on the earth. Section 13.3 highlights the myths that needs to be deflated regarding 5G. Section 13.4 briefs the green technologies to create a more ecofriendly 5G technology and section 13.5 concludes the article.

13.2 Review on Environmental Impacts of 5G

The International Commission on Non-Ionizing Radiation Protection (ICNIRP) is a non governmental body that acts as advisory and guiding board for non-ionizing radiation technologies. The extensive study by the radiation experts reveals that prolonged exposure to microwaves alters the sleep pattern, memory, attention disruption, declined cognitive abilities, increased stress levels, reduced fertility and damaged DNA not only in humans but also in flora and fauna. Many eco systems are disrupted, the effects of which are detrimental on the environment. Figure 13.4 illustrates the adverse impacts of 5G on organisms. The detailed literature presented in this section discusses the adverse effects of 5G network on the flora and fauna present on the earth.

A study on the ecological effects of radio and electromagnetic waves indicated that notable change is observed in the development and reproductive functions of insects and birds [12]. Mammals like mice also exhibited inhibited sperm development [13]. A significant change in the behavior

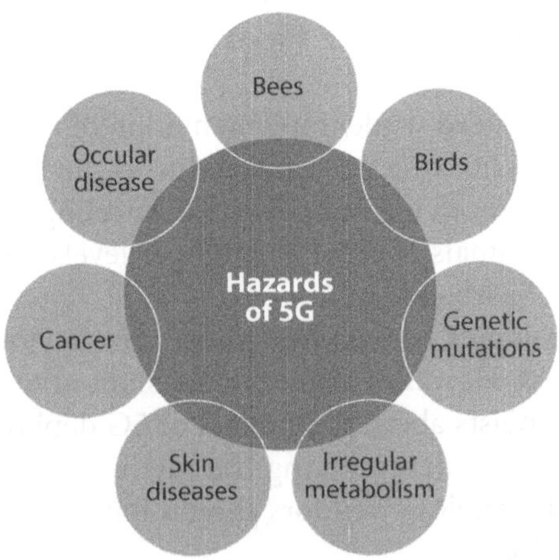

Figure 13.4 Health impacts of 5G on humans and other organisms.

of bees such as decreased reproductivity, reduction in colony strength and dwindled navigation behavior is noted as adverse effect of radiation [14].

The flora on the earth is also damaged due to the radiation effects. The prolonged exposure to RF waves inhibits the production of anthocyanin in trees, thus leading to reduced afforestation rates [15]. This is supported by the study made by Waldmann *et al.*, which reveals that the side of the tree that faces the antenna shows diminished growth pattern [16]. Also, the agricultural crops like peas, maize, fenugreek, onions and tomato are very sensitive and responsive to high frequencies which adversely affects their yield and growth [17]. Prolonged exposure to radiation especially near the BS increases the magnetic property of the materials in proximity. Birds, insects and flora will get adversely affected by these radiations [18].

Humans are also more susceptible to radiation. The pathologists observed that the Code Division Multiple Access (CDMA) and GSM modulated waves led to increase in malignant schwannoma [19]. Subash *et al.* reviewed the potential adverse health effects of 5G in humans. This study presented the impact of 5G technologies in affecting the skin and eyes apart from being carcinogenic [20]. The disruption of DNA, decline in human reproductive capacity and neurodegenerative diseases are also witnessed in humans who are subjected to prolonged exposure of RF waves [1]. Though the radiation from the edge devices are found to be much lower than the base stations, they tend to cause cancer, ocular diseases, metabolism and skin cancer [32]. This forces precautionary measures to be adopted before massive deployment of 5G technologies. Extensive studies have been made in analyzing the genomic instability in elements of human cells. The results reveal that significant changes in development of cell materials are noted in prolonged mobile users [33]. Another notable work in measuring the radiation effect is done by Imtiaz Nasim *et al.* [34]. The experiments indicate that highly intensive EMF is sensed during the downlink of the 5G radio beams. These extremely concentrated energy causes skin effects to the least.

A detailed investigation about the absorption of mm wave in human body in two conditions namely without clothing and clothed is conducted by Gandhi *et al.* The outcome of the experiment portrays that nearly 90-95% of the energy is absorbed, thus the human body starts to act as a impeding transmitter. This will severely damage the cornea of the eye.

The 5G small cells are extremely power thirsty. These technologies are expected to consume approximately thrice the energy than the conventional 4G network. This directly results in depletion of nonrenewable energy generating resources.

The brief review on the impact of 5G technology on human health suggests that the prolonged exposure to RF and EM waves causes heavy damage to flora, fauna and humans on the earth. Though committees like ICNIRP, and Environmental Health Trust (EHT) have devised guideline for radiation levels, it is not practically implemented in many countries across the globe.

13.3 Health Impacts of 5G Technology

Many researchers have highlighted that the Electro Magnetic Field (EMF) disrupts the living condition of all organisms. Federal Communications Commission (FCC) is a regulatory body that monitors the Radio Frequency Radiation (RFR) for any type of wireless data transmissions in United States to be confined within the range [9, 3000] GHz.

FCC regulates the RFR based on the following three metrics and is given in Figure 13.5

i) **Specific Absorption Rate (SAR):** This is the amount of RF absorbed by the human tissue in the near field of the antennas expressed in terms of Watts/Kilogram. SAR values varies based on the transmission frequency, technique for modulation, hardware, reflection and absorption properties of the surrounding materials [23]. The agreeable SAR limit for general public is 1.6W/Kg for 1g tissue cube. Recently,

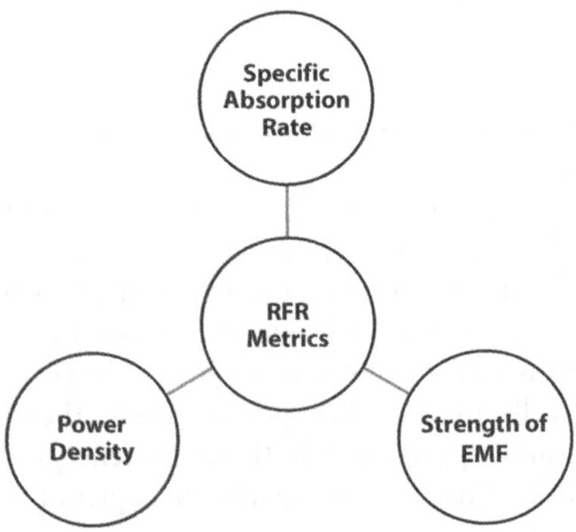

Figure 13.5 RFR metrics as proposed by FCC.

more than a dozen mobile phone models were retired from the market due to their non-compliancy to SAR standard. Studies has revealed that SAR alone is inadequate in deciding the RFR of equipment as they cannot effectively capture the complex interactions and associations between the electromagnetic properties and human tissue.

ii) **Strength of the ElectroMagnetic Field (EMF):** EMF is set up by nature as well as manmade objects. This effect is formed due to the transition of charged particles, which exerts some force on other charged particles. This is generally quantified in Volts/meter. The wireless networks of 3G, 4G and 5G generate radio frequency based electromagnetic fields to transmit data. The antenna gain for Indian BS are set to operate in the range of 15–17 dBi for GSM900 systems and 16–18 dBi for GSM1800 systems [24].

iii) **Power density:** Energy deposition is caused due to the formation of magnetic and electrical fields in objects. This is measured as power density, which is determined from the proximity of the objects to the antenna. The unit of power density is Watts/Sq.m. According to Telecom Regulatory Authority of India (TRAI), the permissible power density for frequencies between 2-300 GHz is 10 Watts/Sq.m.

The regulating bodies monitor the RFR based on the above three metrics. When the emitted RFR do not comply with the standards, then they tend to cause adverse effects on the environment. A very brief classification of the effects of EMF radiation is given in Figure 13.6.

The Bio effects may or may not directly cause harm to humans but they alter the atmospheric conditions. These effects are further bifurcated into

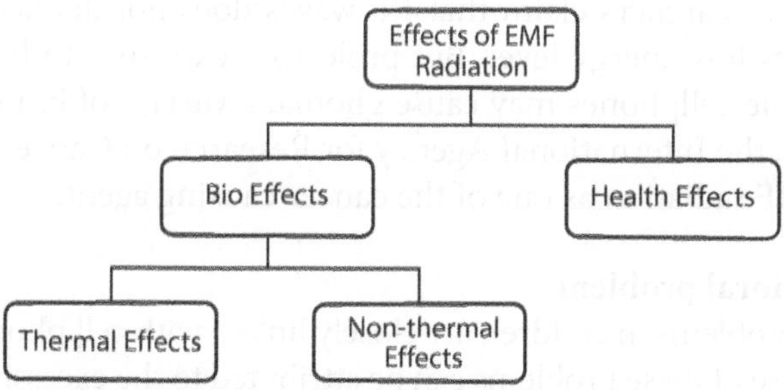

Figure 13.6 Classification of effects of EMF radiation.

thermal and non-thermal effects. Thermal effect increases the temperature in the living tissue which is mainly due to the prolonged exposure to EMF radiation. The non-thermal effects are characterized by the changes in the biological bodies. This may potentially harm the living organisms. The health effects are direct implications sensed in the living organisms due to the EMF exposure. They are certainly harmful to the humans.

The context of developing 5G technologies is to ensure high tech benefits to the mankind. But the environmentalist believe that ICNIRP limits do not protect the organisms and environment. Much stronger limits of radiation is observed in developing nations, which are certainly harmful.

The impulsive detrimental effects caused due to prolonged exposure to EMF radiation are improper fetal development, cancer, behavioral problems, declining of reproductive rate and cognitive dysfunction.

i) Improper fetal development

Women who are exposed to high amount of EMF radiation has elevated chances of miscarriage [25]. The plausible associations between EMF exposure and miscarriages can be linked to the following factors:

1. changes in the permeability of cell membrane of embryos
2. Decreased cellular connections and interactions
3. Abnormal chemical interactions
4. Surge in the free radicals
5. Mitotic malfunctions
6. Permanent damage to cell proteins

Studies reveals that excessive use of cellular devices may lead to spontaneous abortions. At the worst, they may lead to receded embryonic development with shorter gestation period, thus leading to premature delivery.

ii) Cancer

Though the researchers claim that RF waves does not damage the DNA, because of its low energy level. But prolonged exposure to harmful radiation from the cellphones may cause glioma, a variant of brain cancer. To support this, the International Agency for Research on Cancer (IARC) has delineated RF radiation as one of the cancer causing agent.

iii) Behavioral problems

Behavioral problems in children are closely linked with cell phone radiation. The root cause of these problems can be attributed to the excessive radiation.

Children are more likely to suffer from behavioral problems, when they are subjected to abnormal radiation in their mother's womb.

iv) Declined reproductive rate

The EMF radiation has adverse effects on reproductive system. Studies shows that testicular damage can be caused at 0.002 W/kg and worse sperm damage may occur at 0.024 W/kg SAR values [26].

v) Cognitive dysfunction

Scientists has pointed out that cognitive dysfunction is more common in people who are subjected to more cell phone radiation [27]. The EMF affects the Cerebral Blood Flow (CBF) and EEG signals. They also alter the sleep pattern which may cause other issues.

Demystifying the Myths About Environmental Impacts of 5G

It is evident that harmful radiation from 5G technologies alters the behavior and health trends of the living organisms. But recent studies conducted pertaining to the context of environmental impacts of 5G concludes some positive implications.

The first fact is that 5G technologies can lower the Green House Gas (GHG) emissions in the atmosphere. Any modern technology has the 'enabling effect' that is they will be energy efficient. Increasing connectivity and enabling efficient communication will reduce carbon emissions. The deployment of energy efficient power amplifier, smart batteries, user scheduling at BS, tapping renewable energy and more intelligent techniques can reduce the energy consumption without compromising the network QoS. In the user's end 5G technology enables downloading of nearly 5000 ultra-high definition movies for 1KWH against 300 movies in 4G. This double fold energy conservation ascertains the reduced GHG emissions.

A general myth is that 5G is radioactive, which is not true. The BS have to comply with the radiation standards of the country in which the service is hosted. 5G radiation levels confirms that they are not strong enough to cause damage to humans [21].

As 5G deploys low energy millimeter waves, they are not powerful to cause changes in the DNA molecules. Also, the human skin deflects nearly half the amount of power radiated through the high frequency waves.

5G deploys intelligent techniques that makes the devices to be in active mode only when it is required. As a result, 5G devices can drive down the energy, emission levels, e-waste and costs thereby mitigating the effect of climatic change [22]. The video conferencing, transition between workplace and home, improvement in the field of telemedicine during the times of travel restrictions has positively reduced the carbon emissions inside the atmosphere. These are due to the development of high quality network services.

The 5G networks will deploy smaller antennas thus lowering the exposure levels of radiation than the previous generation technologies. This can greatly reduce the negative impact of radiation on living organisms.

13.4 State of Art in Green 5G

Its apparent that 5G technologies demands mass deployment of multitude of technologies in much larger scale. The digital world is now spanning to incubate and station greener network services and segments, monitoring and assessing energy efficiency of mobile networks with innovative strategies to mitigate energy consumption and radiation. Some of the emerging projects to attain the goal of Greener ICT is discussed here and is shown in Figure 13.7.

Energy Aware Radio and neTwork tecHnologies (EARTH)

These are umbrella of solutions aimed to reduce the energy consumption while preserving network QoS [28]. More energy efficient solutions are deployed at various levels namely base stations, radio links, edge devices and radio networks to revolutionize 5G network. EARTH not only proposes energy efficient green solutions, but also integrates them into frameworks, that would suit to development of ICT components. The researchers has categorized the innovations under the canopy of EARTH under three stages:

Figure 13.7 State of art in green 5G.

- Conceptual Stage: Amalgamating the building blocks in accordance with the use cases
- Evaluation Stage: Assessing the configurations of use cases and then integrating them to discern a global solution
- Visualizing Stage: Envisaging the assessment results in a more pedagogic way.

Energy efficient Green networks can be implemented in real time scenarios by choosing appropriate strategies at every stage of their deployment. The parameters such as type and number of active components, traffic variations, net activity level, radio resource allocation technique, network behavior on different time scale are some of the predominant factors that are responsible to achieve energy efficient green 5G under the conception of EARTH.

The propitious tracts for establish the green EARTH solutions are:

- Optimal mix of cell sizes- Achieving good QoS with short Inter site distance in two flavors namely heterogeneous deployment of smaller cells at the edge of relatively bigger macro science and mixture of macro and small cells.
- Relays- Splitting the transmission channel into two parts for better QoS. Two tracks are selected for transmission which may be wither half duplex multi-casting or hybrid relaying scheme by combining Compression-Forward, Decode-Forward and Amplify Forward methods.
- Multi-Radio Access Technology (RAT)- Distribution of available power, choosing active elements and rolling out LTE are the techniques that ensure greener 5G
- Base station cooperation- The issues like poor link quality, poor network coverage and high interference are addressed by base station cooperation by making improvements in cell-edge communication. Fractional Frequency Reuse (FFR) is a much feasible solution to implement base station cooperation.
- Adaptive network reconfiguration- Dynamically modifying the network configurations based on the traffic can greatly reduce the number of active elements.
- Radio resource management- Reducing energy consumption by combining RRM and DTX, multi-cell COMP and traffic dependent multi-RAT coordination based on packet scheduling schemes.

- Future architectures – Scope for incubating more flexibility and intelligence to the systems by replanning the legacy and current days state of art techniques is focused in future architectural frameworks.

Green Meter

Green meter is a significant research conducted by Green Touch to assess the environmental and energy efficiency of entire portfolio of technologies used in communication networks [29]. This comprehensive study considers the depth and breadth of all the architecture, components, edge devices, strategies, algorithms, protocols and technologies deployed in networks. The rampant increase in communication networks in the last decade has accelerated the development of novel technologies and devices, thus greatly increasing the network traffic. But the study reveals that the degree of level of energy consumption has been reduced by the advent of new technologies. The recommendations form this investigation includes adaptation of new ventures and avenues to further contribute to Green network under the umbrella of Mobile access, fixed access and core networks with standardizations:

Mobile Access Networks

- **Green Meter Assessment for Mobile Access Networks**
  This methodology assesses the energy efficiency of mobile networks and thrives as a gold standard for the existing protocols and reference architectures .of the radio networks.
- **Futuristic Base station power model**
  This is a more versatile and realistic power consumption model that could be adapted by all kinds of base stations under all operating conditions. This deploys innovative energy efficiency strategy that could be further tapped for futuristic development.
- **Green Transmission Technologies**
  These technologies are metered in units of bits/second/Watt with good spectrum efficiency. The popular technologies used here are Dynamic MIMO with multi user-multiplexing and inference alignment of antennas
- **Large Scale Antenna Systems (LSAS)**
  Deployment of multiples small, low intensive transmission devices with aggressive multiplexing to render uniform

service across various geographical conditions is ventured by the LSAS technology.

- **Enhanced Wireless Receiver to Mitigate Poor Channel Quality and Interference**
  A restructured receiver model that could function even under low channel quality and high interference is a part of Green meter study. This hardware increases synchronization and iterative channel assessment.

Fixed Access Networks

- **Green Meter Assessment for Fixed Access Networks**
  A fresh architecture is developed that combines multitude of energy efficient technologies in the fixed and metro networks. This model is capable of implementing energy saving strategies even at component level, thud contributing to greater energy efficiency.
- **Cascaded Bit-Interleaving Passive Optical Network (CBI-PON)**
  A comprehensive CBI-PON inculcates the notion of "power :follows the traffic load". This scheme enables data processing to happen at any node.
- **Virtual Home Gateway (vHGW)**
  This is a replacement of the traditional in-house and energy thirsty gateways with gateways with smaller interference. The shared servers are located at cloud, thus facilitating a more generic approach.
- **Point to Point Optical Transceiver**
  This is a novel transceiver that leverages optical technology with more promising data rates. This is realized by embedded innovative hardware design augmented with intelligent power control.

Core Network

- **Green Meter Assessment for Core Networks**
  The energy estimation at core network considers hardware with intelligent control algorithms. This includes a multilayer network optimization framework to evaluate the relative and cumulative energy efficiency mechanisms.

- **Energy Efficient Content Distribution**
  This is a more optimized framework that mitigates energy consumption of content distribution network by scaling the data centers and distributed clouds.

Services, Policies and Standards Activities

The Green Meter study delineates the roles and responsibilities of the working group. The major considerations here are development of traffic volume, policy engagements and contributions to standards.

OPERA-NET2

This is a European project aimed to reduce the adverse effects on the environment due to the network expansion [30]. This is a collaborative venture that affiliates network operators, researchers, academia and industry to invent new models, architectures and design paradigms to mitigate the harmfulness of 5G. The culmination of new technologies happens as a triad as mentioned below:

- **End to end efficiency:**
  This concentrates on manufacturing edge devices and other components especially Remote radio head equipment with recycled plastics and less metals. Cooling mechanisms and liquid cooling prototypes deserve a special mention in implementation of OPERA-NET.
- **Access Network optimization:**
  Improvement in energy optimization by implementing scheduling and energy aware network management in highly dense and heterogeneous networks are primary focus.
- **Optimization of Architecture and change in hardware design:**
  Deploying linear transmitters with promising bandwidth and energy efficiency and optimized power supply architectures are done in OPERA-NET2.

5GrEEn

This is an initiative that concentrates on modifying the architecture, control planes, network construction, radio transmission with backhauling effects to mitigate the adverse effects of the 5G on the environment [31].

This is a combined effort intervened with Mobile and wireless communications Enablers for Twenty-twenty (METIS) to form a European consortium of leading mobile manufacturers, telecommunication operators, industry and academia dedicated to the development of futuristic mobile communications. 5GrEEn will focus on building energy efficient network. Figure 13.8 shows the rampant increase in the mobile traffic between the years 2010-2018.

The primary focus points of the 5GrEEn are:

- **System Architecture**
  The proposed system architecture in 5GrEEn surmise a logical partition among the data transmission and reception. The motivation of this idea emerges from the EARTH project but has ripened in 5GrEE. The cells are created dynamically created to involve in active User Equipment (UE). A cell can be a part of multiple access points with reconfigurable antennae. Hence, the number of access points depend on tne present traffic conditions.
- **Network Deployment**
  The focus is on energy efficient network deployment by adapting different cell topologies, distributed antennas with base stations and other cooperative strategies. Ultra-dense deployments in high intense areas that demands high data rate with minimum power consumption is a characteristic feature of 5GrEEn.
- **Radio Transmission**
  Energy consumption occurs in the cell both during the data transmission and no data transmission. Sensing the direction of the receiver will avoid unnecessary data transmission.

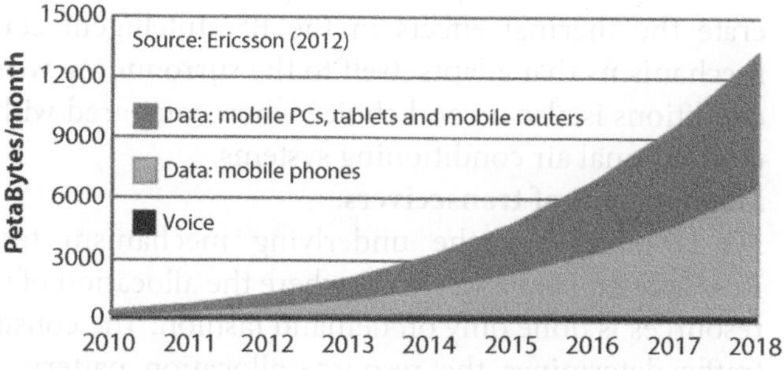

Figure 13.8 Evolution of mobile traffic.

MIMO allows multilayer data transmission that increases the system capacity with much decreased power consumption. Beamforming using multitude of antennas did not explode the power consumption.

- **Backhauling**
  Backhauling is an energy efficient solution in deploying heterogeneous network with increased microwave coverage.

13.5 Deploying Technologies to Achieve Green 5G

Every element in the atmosphere emits and absorbs certain rays. It is not wise to completely defame 5G technology on the environmental grounds. At the same time the service providers and other technological tycoons are striving hard to invent and deploy new technologies to make 5G more clean and green. Some of the notable but more feasible solutions would focus on deploying energy efficient strategies in BS, building energy effective networks, energy aware backhauling, tapping green energy as primary power source to BS, using Artificial Intelligence for power and user management etc. The following are the areas which will positively contribute to the Green 5G:

i) **Energy enhancement and conservation:**
Lots of components are involved in building a BS which requires external power supply. Power amplifier consumes most of the energy and also it dissipates high thermal energy. Replacing the high power consuming amplifiers by double stage configured devices that demands considerably low energy will mitigate the ill effects of power amplifiers.

Also, gravity heat pipe exchanger can be used to moderate the thermal effects in the BS. Intelligent cooling mechanisms that adapts itself to the surrounding weather conditions is also a good choice when compared with the conventional air conditioning systems.

ii) **Management of transceivers:**
5G is built over the underlying mechanism termed Resource on Demand (RoD) where the allocation of radio resources is done only on demand fashion. The consumer traffic determines the resource allocation pattern. This strategy facilitates resource and load sharing where the

unused resources can be turned off. An integrated load and energy sharing strategy will improve network sustainability. Powering the BS through renewable energy will help in offloading of traffic at macrocell and Small Cell Base station by shifting the unused nodes to sleep mode. Intelligent and energy efficient transmission algorithms can be implemented to transit the radio devices to multiple modes such as service, sleep and sniff mode based on its demand.

iii) **Greening the Radio transmissions:**
Since 5G deploys massive MIMO, large number of antenna arrays are embedded in the BS for increasing the efficiency. Cognitive Radio that intelligently detects the communication channel to which the transmissions has to be made can reduce energy consumption without causing interference. Another technology is cooperative relaying in which a third node will act as relay node.

iv) **Establishing renewable energy framework:**
The current BS are powered by nonrenewable energy resources. 5G technologies provisions the use of renewable energy resources like solar, hydro and wind power. Apart from this energy aware configuration and cooperation among cells can improvise the energy efficiency.

Through all the strategies to implant green 5G promises to mitigate the negative effects, they do possess some design challenges:

- The Millimeter-wave technology is totally new and depends on LoS. Hence there are more possibilities for poor coverage.
- The 5G facilitates the usage of green power at the cost of setting the initial infrastructure. Decommissioning and planning the renewable energy management is very tedious task with high capital expenditures.
- Coordinated generation and distribution of renewable energy to the BS must be planned with utmost care, since over sizing and under sizing can cause disruptions.
- The RoD handles the load dynamically, which makes the energy and back up battery planning a difficult task.
- The assured network QoS may fluctuate in extreme weather conditions and under heavy traffic.

It is evident that 5G technologies takes the digital world to next era. These technologies must be implemented with more ecological and environmental responsibility. Various emerging technologies, Artificial intelligence and Green cloud can be explored to make 5G more sustainable.

13.6 Conclusion

The chapter discusses the current 5G technologies with their environmental impacts. It is commonly perceived by the environmentalists that 5G technology can cause severe damage to the entire ecology. This work attempts to demystify the myths in implementing 5G. Also, the plausible solutions with their design challenges for making 5G technology environmentally sustainable are also disused in this article. The researchers can focus on the design considerations to build a green deployable network and make the world a better place to live in.

References

1. Miroslava Karaboytcheva, "Effects of 5G wireless communication on human health," Report by European Parliamentary Research Service, March 2020.
2. Segan, Sascha, "What is 5G?," PC Magazine, January2019.
3. Sharanya S, Revathi Venkataraman, "An intelligent Context Based Multilayered Bayesian Inferential predictive analytic framework for classifying machine states," Journal of Ambient Intelligence and Humanized Computing, https://doi.org/10.1007/s12652-020-02411-2, 2020.
4. Ken Karipidis, Rohan Mate, David Urban, Rick Tinker, Andrew Wood, "5G mobile networks and health—a state-of-the-science review of the research into low-level RF fields above 6 GHz," Journal of Exposure Science & Environmental Epidemiology, https://doi.org/10.1038/s41370-021-00297-6.
5. Wu T, Rappaport TS, Collins CM, "Safe for generations to come: considerations of safety for millimeter waves in wireless communications," IEEE Micro Magazine, 16, pp.65–84, 2015.
6. Gayatri Chittimoju, Usha Devi Yalavarthi, "A Comprehensive Review on Millimeter Waves Applications and Antennas," Journal of Physics: Conference Series, doi:10.1088/1742-6596/1804/1/012205.

7. https://360elsalvador.com/english/what-is-5g-and-what-can-we-expect-from-it/
8. G. Bauch, A. Alexiou, "MIMO technologies for the wireless future," 2008 IEEE 19th International Symposium on Personal, Indoor and Mobile Radio Communications, pp. 1-6, doi: 10.1109/PIMRC.2008.4699969, 2008.
9. Mostafa Zaman Chowdhury, Takeo Fujii, Gabriel-Miro Muntean, Ji-Woong Choi, Giuseppe Araniti, "Emerging Small Cell Wireless Technologies for 5G: Architectures and Applications," Emerging Small Cell Wireless Technologies for 5G: Architectures and Applications, https://doi.org/10.1155/2018/8969264.
10. F. Alavi, K. Cumanan, Z. Ding, A. G. Burr, "Beamforming Techniques for Nonorthogonal Multiple Access in 5G Cellular Networks," IEEE Transactions on Vehicular Technology, vol. 67, no. 10, pp. 9474-9487, Oct. 2018, doi: 10.1109/TVT.2018.2856375.
11. Z. Zhang, X. Chai, K. Long, A. V. Vasilakos, L. Hanzo, "Full duplex techniques for 5G networks: self-interference cancellation, protocol design, and relay selection," IEEE Communications Magazine, vol. 53, no. 5, pp. 128-137, 2015.
12. Cucurachi S, Tamis WL, Vijver MG, Peijnenburg WJ, Bolte JF, de Snoo GR," A review of the ecological effects of radiofrequency electromagnetic fields (RF-EMF),". Environ Int,,51, pp.116-40, 2015.
13. Otitoloju AA, Obe IA, Adewale OA, Otubanjo OA, Osunkalu VO, "Preliminary study on the induction of sperm head abnormalities in mice, Mus musculus, exposed to radiofrequency radiations from global system for mobile communication base stations," Bull Environ Contam Toxicol, 84(1), pp: 51-4, 2010.
14. Favre, D.," Mobile phone-induced honeybee worker piping," Apidologie, 42, 270–279, 2011.
15. Katie Haggerty, "Adverse Influence of Radio Frequency Background on Trembling Aspen Seedlings: Preliminary Observations," International Journal of Forestry Research, 2010.
16. Cornelia Waldmann-Selsam, Alfonso Balmori-de la Puente, Helmut Breunig, Alfonso Balmori, "Radiofrequency radiation injures trees around mobile phone base stations, Science of The Total Environment," 572, pp. 554-569, 2016.
17. Halgamuge MN, "Review: Weak radiofrequency radiation exposure from mobile phone radiation on plants," Electromagn Biol Med, 36(2), pp. 213-235, 2017.

18. Alfonso Balmori, "Anthropogenic radiofrequency electromagnetic fields as an emerging threat to wildlife orientation," Science of The Total Environment, 518–519, pp. 58-60, 2015.
19. J. C. Lin, "Clear Evidence of Cell Phone RF Radiation Cancer Risk [Health Matters]," IEEE Microwave Magazine, 19 (6), pp. 16-24, 2018.
20. Subhash Chander Verma, Tejaswini T M, Devasis Pradhan, "Harmful Effects Of 5G Radiations: Review," International Journal of Industrial Electronics and Electrical Engineering, 7(3), 2019.
21. https://www.theguardian.com/technology/2020/mar/12/5g-safe-radiation-watchdog-health
22. https://www.ericsson.com/en/blog/2021/1/digitalization-5g-climate-action.
23. National Collaborating Centre for Environmental Health, British Columbia Centre for Disease Control, radiofrequency toolkit for environmental health practitioners, Vancouver, BC, Canada, http://www.bccdc.ca/resource-gallery/Documents/Educational%20Materials/EH/Radiofrequency-Toolkit.pdf, (2014) accessed July 5, 2019.
24. "Information paper On Effects of Electromagnetic Field Radiation from Mobile Towers and Handsets," TRAI, 2014.
25. Abad, M., Malekafzali, H., Simbar, M., Seyed Mosaavi, H., Merghati Khoei, E," Association between electromagnetic field exposure and abortion in pregnant women living in Tehran," International journal of reproductive biomedicine, 14(5), 347–354, 2016.
26. S. Dasdag, M. Taş, M.Z. Akdag, K. Yegin, "Effect of long-term exposure of 2.4 GHz radiofrequency radiation emitted from Wi-Fi equipment on testes functions," Electromagn. Biol. Med, 34, pp. 37–42, 2015.
27. Shubhra Jain, "Effect of Cell Phone Radiation on Neurological Function," Submitted as coursework for PH250, Stanford University, Spring 2012.
28. EARTH project deliverable, "Final integrated concept," Jul. 2012.
29. GreenTouch Foundation, "GreenTouch final results from Green Meter research study," White Paper V1.0, Jun. 2015.
30. Lasanen, M., Aubree, M., Cassan, C., Conte, A., David, J., Elayoubi, S. E., ... & Smalen, L. (2013, May). Environmental friendly mobile radio networks: Approaches of the European OPERA-Net 2 project. In *ICT 2013* (pp. 1-5). IEEE.
31. Olsson, M., Cavdar, C., Frenger, P., Tombaz, S., Sabella, D., & Jantti, R. (2013, October). 5GrEEn: Towards Green 5G mobile networks. In *2013 IEEE 9th international conference on wireless and mobile*

computing, networking and communications (WiMob) (pp. 212-216). IEEE.

32. L. Chiaraviglio, A. Elzanaty and M. -S. Alouini, "Health Risks Associated With 5G Exposure: A View From the Communications Engineering Perspective," in *IEEE Open Journal of the Communications Society*, vol. 2, pp. 2131-2179, 2021, doi: 10.1109/OJCOMS.2021.3106052.
33. Gandhi G, Singh P, Kaur G. Perspectives Revisited - The Buccal Cytome Assay in Mobile Phone Users. International Journal of Human Genetics. 2017;15(4):173-82.
34. I. Nasim and S. Kim, "Adverse Impacts of 5G Downlinks on Human Body," *2019 SoutheastCon*, 2019, pp. 1-6, doi: 10.1109/SoutheastCon42311.2019.9020454.
35. O. P. Gandhi and A. Riazi, "Absorption of Millimeter Waves by Human Beings and its Biological Implications," in *IEEE Transactions on Microwave Theory and Techniques*, vol. 34, no. 2, pp. 228-235, Feb 1986, doi: 10.1109/TMTT.1986.1133316.

14

Enhanced Human Growth Through Technical Financial Literacy

R. Muthu Lakshmi[1] and A. Maria Bharathi[2*]

[1]*Sri Sarada College for Women (Autonomous), Tamil Nadu, India*
[2]*Department of Commerce, Sri Sankara Bhagavathi Arts and Science College, Kommadikottai, Tamil Nadu, India*

Abstract
Financial literacy plays an important role to develop financial inclusion and growth. Growth is needed for every human being in education, the standard of living, knowledge etc., and financial literacy is the first step to achieving growth. Technical financial literacy is likely to become a progressively important aspect of education. The growth of the economy means that people will become more accountable for their own monetary planning. In the modern world, people will need to manage their funds individually with available financial knowledge concerned with savings and investment plans. Financial literacy is significant because it has been connected to commercial behaviour and wealth accumulation of individuals by contributing to the profitable growth and development of the country. Thus financial literacy will show the main role in understanding the complexity of the financial market progress of new economic products and changes in economic factors. Financial literacy programs will improve growth in all the activities among people, students and Investors.

Keywords: Financial inclusion, financial literacy, bank, post office and insurance

**Corresponding author*: bharathiduraipandi23@gmail.com

A. Suresh, J. Ramkumar, M. Baskar and Ali Kashif Bashir (eds.) Resource Management in Advanced Wireless Networks, (291–304)

14.1 Introduction

Financial literacy allows the public person and thus diminishes the load of caring the common person from the fundamentals of market disappointment, attributable to, de facto, information asymmetries. For example, the importance on market discipline, as one of the three supports of banking guideline, particularly under Base I and II, and is best served by contribution of financially knowledgeable bank customers in the financial market place. Financial literacy not only will recover the quality of life rather it can improve the reliability and quality of markets. It can encourage saving and thus, confirm that investors can enjoy a distinguished life after retirement through correct monetary planning and investment. Financially cultivated consumers, in turn, can advantage the economy by boosting genuine rivalry, forcing the service suppliers to modernise and improve their levels of efficacy. Financial literacy will improve investors' thoughtful of financial markets, products, ideas and risk connected with the same.

Financial literacy is significant because it has been associated to economic performance and wealth build-up of individuals by contributing to economic progress and improvement of the country.

14.2 Statement of the Problem

Poverty is a well-known problem in most of the developing and underdeveloped countries. The inadequate educational status, poor literacy and unawareness about the financial stream have been a reason for it. Members of scheduled communities are poor in literacy, which makes them further away from others. Hence they do not use formal financial sources for their requirements.

Financial literacy has believed greater importance in latest years. It is popularly assumed that people require to become more self-dependent in the upcoming years. Growing contest has strained service providers to present hybrid products with composite characteristics. Many users have left these ill-equipped products to cope with the difficult choices which they need to make due to inadequate alertness and inappropriate consciousness level.

14.3 Objectives of the Study

- To know the attentiveness about finance among Scheduled Communities in SathankulamTaluk.
- To find out the saving habits of the people.

To give suggestions to improve saving habits among Scheduled Communities in SathankulamTaluk

14.4 Review of Literature

MahikaChhatwani and Sushanta Kumar Mishra (2021)[1] in their study entitled **"Financial fragility and financial optimism linkage during COVID-19: Does financial literacy matter?"** shows that COVID-19 and its significance convinced many contests for persons and many of them well-informed in financial instability. Monetary positivity is essential in this condition as it assistances persons and establishments to get better from such circumstance. Under this study experienced if fiscal literacy can diminish financial instability's unfavourable effect on financial self-assurance. Findings of the study reveals that a pessimistic relationship between financial weakness and fiscal positivity. The association was durable for women. To speech probable endogeneity, the study conducted fitness analysis using active variable regression and propensity score matching. The findings of the study provide suggestions to upturn financial assurance during the endemic.

Georgios A. Panos & John O.S Wilson (2020)[2] in their study on **"Financial Literacy and responsible finance in the Fin Tech era: capabilities and challenges"** suggested that fiscal mastery acting an very vital element in financial existence, and that variances in economic awareness attained most key in life can clarify a considerable part of monetary and additional in common well-being in grown life. The study cover four expansive subjects correlated to monetaristability, namely Fin Tech apps and fiscal ability, financial inclusion and pecuniary welfare, student commercial literacy and monetary grades and information, fiscal direction and fraud invention. The conclusions show that fraud strategies are progressively more not simple and it is upper financial knowledge relatively than fundamental money administration skills that delivers the degree of difficulty necessary to reduce fraud. The paper attractions policy inferences for purchaser education programmes to go away from illuminating money administration skills, and provides bigger financial knowledge necessary for following fraud.

---

[1]MahikaChhatwani, Sushanta Kumar Mishra (2021). Financial fragility and financial optimism linkage during COVID-19: Does financial literacy matter?. Journal of Behavioral and Experimental Economics. Vol.94.

[2]Georgios, A. Panos & John, O.S Wilson (2020). Financial Literacy and responsible finance in the Fin Tech era: capabilities and challenges. The European Journal of Finance, Vol.26, Issue.4-5.

ThiAnhNhu Nguyen and Kieu Minh Nguyen (2020)[3] in their study entitled **"Role of financial literacy and peer effect in promotion of financial market participation: Empirical Evidence in Vietnam"** examines that how communications of pecuniary literacy and examine result pointers persuade financial marketplace contributions of Vietnamese investors. The result states that persons with superior level of financial literacy, surely those with progressive pecuniary literacy level have a propensity to contribute in financial marketplace. Fundamental pecuniary literacy level is measured universal facts linked to daily process of individuals, while progressive financial literacy level emphases more on complicated data linked to dangerous asset matter and savings presentation. The result illustrates that highly developed fiscal literacy level and professed and statistically important force on human being monetary market sharing, whereas crucial financial literacy has a major unnecessary effect on individual financial bazaar involvement. It advises that policy maker have to make and expand monetary literacy programs so that people can obtain improved thoughtful of pecuniary acquaintance and become more secure to receive ingredient in financial marketplace.

Anshika (2017)[4] in the study on **"Financial Literacy – A Present Day Need in India"** explained that financial literacy has secures a vital role in the recent scenario due to many causes viz. progress of the new financial products, complication of the financial markets facts asymmetry and the variations in the other commercial factors in the economy. The literacy level of India is very low as compared to BRICS nations and European nations in the year 2015. Lack of financial literacy leads to the : 11- informed financial decisions and these choices in tam, have marvellous undesirable influence on the financial system as a whole the present study aims to examine the various creativities taken by government to recover the level of financial literacy in Indian.

Henager, Robin and Cude, Brenda J. (2016)[5] in their study on **"Financial Literacy and Long and Short-Term Financial Behavior in Different Age Groups"** examine the association between literacy and financial behaviours among several age groups. Under this study financial

[3]ThiAnhNhu Nguyen and Kieu Minh Nguyen (2020). Role of financial literacy and peer effect in promotion of financial market participation: Empirical Evidence in Vietnam. The Journal of Asian Finance, Economics and Business, Vol. 7, No.6.

[4]Anshika (2017). Financial Literacy – A Present Day Need in India. Jaipuria International Journal of Management Research, Vol. 3, No.1.

[5]Henager, Robin; Cude, Brenda J. (2016). Financial Literacy and Long and Short-Term Financial Behavior in Different Age Groups. Journal of Financial Counselling and Planning, Vol. 27.

literacy is measured in three this ways: objective financial knowledge, subjective financial knowledge or confidence, and subjective financial management ability. Long-term monetary behaviour referred to retirement saving and investing behaviour, whereas short-term monetary performance referred to spending and emergency saving behaviour. The age groups were 18-24, 25-32, 35-44, 45-54, 55-64 and 65 and Older. The findings of the study is that, in the full sample, both objective and subjective financial literacy variables were absolutely connected with long and short term financial behaviour. In the age sub samples, subjective financial knowledge or confidence way more strongly related to long and short term financial behaviour than either objective financial knowledge or subjective financial administration aptitude in the older age groups, objective financial knowledge is more powerfully related to long – term financial behaviour than either of the other two measures of financial literacy.

Eniola, Anthony Abiodun and Entebang, Harry (2016)[6] in their study on **"Financial literacy and SME firm performance"** lead the research based on resource based view theory to show the effect of financial literacy on the performance of small and medium enterprises (SMEs) in Nigeria. This study argues one of the important issues: financial literacy is a exacting trouble of small and medium enterprises (SMEs), company be inclined to have much developed charge of occupation enlargement, but also are more likely to go out of business or stay behind student due to institutional and financial literacy struggle. The connotation of the small and medium enterprise (SMEs) sector is documented in economics worldwide, irrespective of the economy's advance stage. The involvement on route presentation expansion, employ creation and social progress is pleasant greatly and small and medium enterprises (SMEs) reflect as a key factor in a successful formula for attain economic growth. Subsequently, financial literacy is now internationally standard as a major factor of economic and financial constancy and improvement of SMEs for performance.

SinemAtes Ali Coskun, M. Abdullah Sahin, M. and LeventDemircan (2016)[7] in their study on **"Impact of Financial Literacy on the Behavioral Biases of Individual Stock Investors: Evidence from Borsa Istanbul"** examine the connection among the financial literateness of investor and the point of behavioural favouritisms that may result in unreasonable

---

[6]Eniola, Anthony Abiodun and Entebang, Harry (2016). Financial literacy and SME firm performance. International Journal of Research Studies in Management, Vol.5, No.1.

[7]SinemAtes Ali Coskun, M. Abdullah Sahin, M. LeventDemircan (2016). Impact of Financial Literacy on the Behavioral Biases of Individual Stock Investors: Evidence from Borsa Istanbul. Business and Economics Research Journal, Vol.7, No.3.

behaviour in economic judgement creation. And also scrutinize the demographic influences that affect the point of financial literacy and behavioural biases. Results recommend that in the order of half of the investors have a small financial literacy level their main source of fiscal information is recommendation from parents or friends and they have a high level of behavioural biases. While some of this behavioural bias is autonomous of the level of financial literacy there is an important connection between a number of other biases and the level of financial literacy.

MustabsarAwais, M, FahadLaber, Nilofer Rasheed and Aisha Khursheed (2016)[8] in their study on **"Impact of Financial Literacy and Investment Experience on Risk Tolerance and Investment Decisions: Empirical Evidence from Pakistan"** tries to build a theoretical model that shows to be useful for this investors while making the security or designing portfolio. During the period of investment decision making, investors meeting to highly complex factors such as risk, vagueness, and choice over load. Investors have to class risks in their financial decisions. Greater investment involvements and financial literacy will lead to bigger risk tolerance and investor then have to select risky investment securities to match with their high level of risk tolerance knowledgeable investors have a portfolio of worthy and bad experience. By the bigger level of knowledge about financial information and the increased capability of investigating that information, investor can recover the capacity jump into dangerous investments for receiving high returns by handling investment efficiently. The research proves to be appreciated as it has discovered this impact as well. Financial literacy which is also connected with the awareness of factors such as price rises, price decreases compounding interest and diversification of risk, as the examination of these factors also helps to go for appropriate choices.

VlastaBahoec, DajanaBarbic and Irena Palic (2015)[9] in their study on **"Testing the effects of financial literacy on debt behavior of financial consumers using multivariate analysis methods"** the aim of this study is three hold. First, to detect and categorize formation financial consumers financial literacy and analyse whether the different levels of financial literacy is

[8]MustabsarAwais, M,FahadLaber, Nilofer Rasheed, Aisha Khursheed (2016). Impact of Financial Literacy and Investment Experience on Risk Tolerance and Investment Decisions: Empirical Evidence from Pakistan. International Journal of Economics and Financial Issues, Vol.6, No.1.

[9]VlastaBahoec, DajanaBarbic, Irena Palic (2015). Testing the effects of financial literacy on debt behavior of financial consumers using multivariate analysis methods. CRORR Journal, Vol.6, No.2.

statistically important in terms of socio-demographic features, the second objective is determining whether a respondents debt behaviour differs based on their financial literacy. Further, the paper aims to inspectbases of different debt behaviour in relative to monetary literacy. The result suggests that respondents display different debt behaviour due to different levels of financial literacy. In addition, the level of throwaway house hold income per house hold member was originate to be statistically unimportant with respect to different levels of financial literacy, which such levels were found to be statistically significantly dissimilar when matched to a respondent's gender.

14.5 Sampling Design

Under this study, primary data is collected by the researcher through convenience sampling method. Since the numbers of respondents are in several and the study period is limited, convenience sampling is employed. However, the 150 samples are selected from SathakulamTaluk. Appropriate representation is given for rural and urban respondents. Further, the respondents selected belong to different income groups.

14.6 Methodology

The prime data were composed from respondents with the carry of interview schedule. The interview schedule was constructed by the researcher to collect information about financial literacy by visiting people in SathankulamTaluk from 150 respondents has been selected by convenience sampling method.

14.7 Financial Literacy Among Scheduled Community

Financial awareness is necessary for everyone in the society particularly for the scheduled community to access bank, postal, insurance and other financial services. The level of financial literacy among scheduled community people in SathankulamTaluk is presented in Table 14.1.

It is clear from the table that out of 150 respondents in the study 96 respondents (64 per cent) who are financially literates have scored between 1 to 5 and 38 respondents (25 percent) who have the financial literacy scored below 1; and 16 respondents (11 percent) who have financial literacy scored 5 to 10.

In order to measure financial literacy in relation to these eight elements, the above 50 questions were asked. And from the answers, the financial

Table 14.1 Financial literacy among scheduled community.

Sl. no	Financial literacy scores	No. of respondents	Percentage (%)
1	below 1	38	25
2.	1 to 5	96	64
3	5 to 10	16	11
	Total	**150**	**100**

Source: Primary data.

literacy level of the respondents in relation to the above eight elements is measured, and the details are furnished below.

14.8 Financial Literacy on Investments

The surplus fund with individuals in invested in land, bank, or post office, financial literacy on investment score secured by the scheduled community people in the study area is presented in Table 14.2.

Table shows that 97 respondents (65 percent) who have the financial literacy on investment have scored between 1 to 2 and 53 respondents (35 per cent) who have financial literacy on investment have scored below 1.

14.9 Financial Literacy on Borrowings

Borrowing means receiving fund from a lender or investor with the accepting that the full amount will be repay in the future, generally with interest. Financial literacy on borrowings score secured by the scheduled community people in the study area is presented in Table 14.3.

Table 14.2 Financial literacy on investments.

Sl. no	Financial literacy scores	No. of respondents	Percentage(%)
1	below 1	53	35
2	1 to 2	97	65
	Total	**150**	**100**

Source: Primary data.

Table 14.3 Financial literacy on borrowing.

Sl. no.	Financial literacy scores	No. of respondents	Percentage (%)
1	below 1	83	55
2	1 to 2	67	45
	Total	**150**	**100**

Source: Primary data.

From the table shows in clear that 83 respondents (55 percent) who have the financial literacy on borrowing have scored between 1; and 67 respondents (45 per cent) who have financial literacy on borrowing have scored below 1 to 2.

14.10 Financial Literacy on Debt Management

A debt is created when a creditor lends a sum of assets to a debtor Financial literacy on debt management score secured by the scheduled community people in the study area is presented in Table 14.4.

Table reveals that 116 respondents (77 percent) who have the financial literacy on debt management have scored between 1; and 34 respondents (23 per cent) who have financial literacy on debt management have scored below 1 to 2.

14.11 Financial Literacy on Debit and Credit Cards

Debit and credit cards are used to create an electronic withdrawal from funds on deposits in a bank as in purchasing goods or obtaining cash advances. Financial literacy on debit and credit cards score secured by the scheduled community people in the study area is presented in Table 14.5.

Table 14.4 Financial literacy on debt management.

Sl. no.	Financial literacy scores	No. of respondents	Percentage (%)
1	below 1	116	77
2	1 to 2	34	23
	Total	**150**	**100**

Source: Primary data.

Table 14.5 Financial literacy on debit and credit cards.

Sl. no.	Financial literacy scores	No. of respondents	Percentage (%)
1	below 1	101	67
2	1 to 2	49	33
	Total	150	100

Source: Primary data.

Table highlights that out of the total 101 respondents (67 percent) who have the financial literacy on debit and credit cards have scored between 1; and 49 respondents (33 per cent) who have financial literacy on debit and credit cards have scored below 1 to 2.

14.12 Financial Literacy on Capital Market

People may invest the surplus funds in shares and debentures. Financial literacy on capital market score secured by scheduled community people in the study are is presented in Table 14.6.

Table shows that out of the total 131 respondents (87 percent) who have the financial literacy on capital market have scored between 1; and 19 respondents (13 per cent) who have financial literacy on capital market have scored below 1 to 2.

14.13 Financial Literacy on Insurance

Life is full of risk and uncertainty. Insurance is an invaluable means to provide protection against risks. Literacy on insurance draws more people to take insurance policy and be financially included. Financial literacy on insurance score secured by the scheduled community people in the study area is presented in Table 14.7.

Table 14.6 Financial literacy on capital market.

Sl. no.	Financial literacy scores	No. of respondents	Percentage (%)
1	below 1	131	87
2	1 to 2	19	13
	Total	150	100

Source: Primary data.

Table 14.7 Financial literacy on insurance.

Sl. no	Financial literacy scores	No. of respondents	Percentage (%)
1	below 1	102	68
2	1 to 2	48	32
	Total	**150**	**100**

Source: Primary data.

From the table it is clear that out of the total of 150 respondents, 102 respondents (68 percent) who have financial literacy on insurance have scored below 1; and 48 respondents (32 per cent) who have financial literacy on insurance have scored between 1 to 2. It shows the poor literacy level in scheduled communities in SathankulamTaluk.

14.14 Financial Literacy on Postal Services

The post offices give monetary services provide to the different needs of the community. Financial literacy on postal services score protected by scheduled community people in the study area is presented in Table 14.8.

Table reveals that out of the total 101 respondents (67 percent) who have the financial literacy on postal services have scored between 1; and 49 respondents (33 per cent) who have financial literacy on postal services have scored below 1 to 2.

14.15 Financial Literacy on Other Financial Services

Financial literacy on other financial services score secured by scheduled community people in the study area is presented in Table 14.9.

Table 14.8 Financial literacy on postal services.

Sl. no	Financial literacy scores	No. of respondents	Percentage (%)
1	below 1	101	67
2	1 to 2	49	33
	Total	**150**	**100**

Source: Primary data.

Table 14.9 Financial literacyon other financial services.

Sl. no	Financial literacy scores	No. of respondents	Percentage (%)
1	Below 1	105	70
2	1 to 2	45	30
	Total	**150**	**100**

Source: Primary data.

Table shows that 105 respondents (70 percent) who have the financial literacy on other financial services have scored between 1; and 45 respondents (30 per cent) who have financial literacy on other financial services have scored below 1 to 2.

14.16 Conclusion

The study measures the levels of financial awareness among scheduled communities in SathankulamTaluk, through the 8 elements of financial awareness. It is found in the study that 65 percent of the respondents are financially literates through investment, 45 per cent of the respondents are economically literates in the course of borrowings. Another 23 percent of the respondents are financially literates through debt management and 33 percent of the respondents are financially literates through debit and credit card. Further it is identified that 13 percent of the respondents are financially literates through capital market. As a whole, it may be concluded that the members of scheduled community are financially liberated at a lower level in SathankulamTaluk.

The Government may takes many steps to recover the level of monetary inclusion and monetary literacy among scheduled communities during this pandemic period. To recover the financial literacy level, Government must take steps like awareness programmes and digital financial literacy programmes among scheduled communities.

14.17 Bibliography

14.17.1 Books

- Cooper, D.R. & Schindler, P.S. (2003). Business Research Methods. New Delhi: Tata McGraw-Hill.
- Fisher, R., (1938). Statistical Methods for Research Workers. Edinburgh: Oliver and Boyd.
- G.C,H., (1970). Research Concepts in Human Behaviour Education, Psychology, Sociology. New York: Appleton-century crofts, p.448.
- Kothari, C. and Garg, G., 2014. Research Methodology Methods and Techniques. 3rd ed. New Delhi: New Age International (P) Ltd., p.63.
- Krishnasamy, O.R. and Ranganatham, M (2018). Research methodology for social science research. Himalaya publications, New Delhi.

14.17.2 Journals

- Abraham Ansong and Michael AsieduGyensare (2012). Determinants of University Working-Students' Financial literacy at the University of Cape Coast, Ghana. International Journal of Business and Management, Vol.7, No.9
- Adriana Zait and Patricea Elena Bertea (2014). Financial Literacy – Conceptual Definition and Proposed Approach for a Measurement Instrument. The Journal of Accounting and Management, Vol.4, No.3
- Adriana Zait and Patricea Elena Bertea (2015). Financial Literacy – Conceptual Definition and Proposed Approach for a Measurement Instrument. The Journal of Accounting and Management, Vol.4, No.3
- Ahmed Sameer E LKhatib (2021). Financial Literacy. International Journal for Innovation Education and Research, Vol.9,No.3
- AinaCaplinska, Alina Ohotina (2019). Analysis of Financial Literacy Tendencies with Young People. The Journal of Socio Sustainability, Vol.6, No.4
- Alan L. Gustman Thomas L. SteinmeierNahidTabatabai (2012). Financial Knowledge and Financial Literacy at the

Household Level. American Economic Review, Vol.102, No.3

- Alejandro Drexler, Greg Fischer and Antoinette Schoar (2014). Keeping it Simple: Financial Literacy and Rules of Thumb. American Economic Journal: Applied Economics, Vol.6, No.2
- Atkinson, A., & Messy, F. (2012). Measuring financial literacy: Results of the OECD/International Network on Financial Education (INFE) pilot study (OECD Working Papers on Finance, Insurance and Private Pensions, No. 15). Paris: OECD Publishing. DOI: http://dx.doi.org/10.1787/5k9csfs90fr4-en.
- Kiliyanni, A.L., & Sivaraman, S. (2016). The perception – reality gap in financial literacy: Evidence form the most literate state in India. International Review of Economics Education, 23, 47 – 64.
- Jerold Arockia, V. (2008), "Financial Inclusion and literacy in India," Indian Journal of Finance, Vol.2, N0. 3, July.
- M. Ramji, Financial Inclusion in Gulbarga: Finding Usage in Access, Chennai; IFMR, 2009.
- Mitra Rona (2006), "Financial Inclusion: Meeting the challenge", people's Democracy, April 22 & 29, 2007.
- UshaThorat (2007), 'Financial Inclusion, The Indian Experience', Text of the speech delivered at the H MT-DFID Financial Inclusion Conference White hall place, London, June 19, 2007, RBI, Mumbai.

About the Editors

A. Suresh, PhD, is an associate professor in the Department of Computer Science and Engineering at SRM Institute of Science and Technology, Tamil Nadu, India. With almost two decades of experience, he has published three patents and 90 papers in technical journals. He has also authored several books and is currently writing a number of other books. He has published more than 40 papers at international scientific conferences, and he has served as editor and reviewer for various journals.

J. Ramkumar, PhD, is an assistant professor in the Department of Computer Science and Engineering, School of Computing, SRM Institute of Science and Technology, Chennai, India. His research interests include broadband wireless networks and scheduling in WiMAX and LTE.

M. Baskar, PhD, is a researcher and faculty member in the Department of Computer Science and Engineering, School of Computing, Kattankulathur Campus, SRM Institute of Science and Technology. He has published 35 research articles in scientific journals and 13 papers at international conferences. He is a reviewer for one scholarly journal.

Ali Kashif Bashir, PhD, is a faculty member of the School of Computing and Mathematics, Manchester Metropolitan University, United Kingdom. He is an adjunct professor at the School of Electrical Engineering and Computer Science, National University of Science and Technology, Islamabad. He is the author of over 90 peer-reviewed articles and is serving as the editor-in-chief of the "IEEE Future Directions Newsletter." He is the editor of several journals and has served as a guest editor on several special issues in journals. He has served as chair on several conferences and workshops.

Index

Also of Interest

MODELING AND OPTIMIZATION OF OPTICAL COMMUNICATION NETWORKS Edited by Chandra Singh, Rathishchandra R Gatti, K.V.S.S.S.S. Sairam, and Ashish Singh, ISBN: 9781119839200. *Modeling and Optimization of Optical Communication Networks* is a comprehensive and authoritative book that delves into the optical networks, principles, technologies, and practical applications of optical networks, equipping readers with the knowledge needed to design, implement, and optimize optical networks for various applications, from telecommunications to scientific research. With its comprehensive coverage and up-to-date insights, this book serves as an essential reference in the field of optical networks.

RF CIRCUITS FOR 5G APPLICATIONS: Designing with mm Wave Circuitry, Edited by Sangeeta Singh, Rajeev Kumar Arya, B.C. Sahana and Ajay Kumar Vyas, ISBN: 9781119791928. This book addresses FinFET-based analog IC designing for fifth generation (5G) communication networks and highlights the latest advances, problems, and challenges while presenting the latest research results in the field of mmwave integrated circuits design.

WIRELESS COMMUNICATION SECURITY: Mobile and Network Security Protocols, Edited by Manju Khari, Manisha Bharti, and M. Niranjanamurthy, ISBN: 9781119777144. Presenting the concepts and advances of wireless communication security, this volume, written and edited by a global team of experts, also goes into the practical applications for the engineer, student, and other industry professionals.

ADVANCES IN DATA SCIENCE AND ANALYTICS, Edited by M. Niranjanamurthy, Hemant Kumar Gianey, and Amir H. Gandomi, ISBN: 9781119791881. Presenting the concepts and advances of data science and analytics, this volume, written and edited by a global team of experts, also goes into the practical applications that can be utilized across multiple disciplines and industries, for both the engineer and the student, focusing on machining learning, big data, business intelligence, and analytics.

ARTIFICIAL INTELLIGENCE AND DATA MINING IN SECURITY FRAMEWORKS, Edited by Neeraj Bhargava, Ritu Bhargava, Pramod Singh Rathore, and Rashmi Agrawal, ISBN 9781119760405. Written and edited by a team of experts in the field, this outstanding new volume offers solutions to the problems of security, outlining the concepts behind allowing computers to learn from experience and understand the world in terms of a hierarchy of concepts.

MACHINE LEARNING AND DATA SCIENCE: Fundamentals and Applications, Edited by Prateek Agrawal, Charu Gupta, Anand Sharma, Vishu Madaan, and Nisheeth Joshi, ISBN: 9781119775614. Written and edited by a team of experts in the field, this collection of papers reflects the most up-to-date and comprehensive current state of machine learning and data science for industry, government, and academia.

MEDICAL IMAGING, Edited by H. S. Sanjay, and M. Niranjanamurthy, ISBN: 9781119785392. Written and edited by a team of experts in the field, this is the most comprehensive and up-to-date study of and reference for the practical applications of medical imaging for engineers, scientists, students, and medical professionals.

SECURITY ISSUES AND PRIVACY CONCERNS IN INDUSTRY 4.0 APPLICATIONS, Edited by Shibin David, R. S. Anand, V. Jeyakrishnan, and M. Niranjanamurthy, ISBN: 9781119775621. Written and edited by a team of international experts, this is the most comprehensive and up-to-date coverage of the security and privacy issues surrounding Industry 4.0 applications, a must-have for any library.

CYBER SECURITY AND DIGITAL FORENSICS: Challenges and Future Trends, Edited by Mangesh M. Ghonge, Sabyasachi Pramanik, Ramchandra Mangrulkar, and Dac-Nhuong Le, ISBN: 9781119795636. Written and edited by a team of world renowned experts in the field, this groundbreaking new volume covers key technical topics and gives readers a comprehensive understanding of the latest research findings in cyber security and digital forensics.

DEEP LEARNING APPROACHES TO CLOUD SECURITY, Edited by Pramod Singh Rathore, Vishal Dutt, Rashmi Agrawal, Satya Murthy Sasubilli, and Srinivasa Rao Swarna, ISBN 9781119760528. Covering one of the most important subjects to our society today, this editorial team delves into solutions taken from evolving deep learning approaches, solutions allow computers to learn from experience and understand the world in terms of a hierarchy of concepts.

MACHINE LEARNING TECHNIQUES AND ANALYTICS FOR CLOUD SECURITY, Edited by Rajdeep Chakraborty, Anupam Ghosh and Jyotsna Kumar Mandal, ISBN: 9781119762256. This book covers new methods, surveys, case studies, and policy with almost all machine learning techniques and analytics for cloud security solutions.

SECURITY DESIGNS FOR THE CLOUD, IOT AND SOCIAL NETWORKING, Edited by Dac-Nhuong Le, Chintin Bhatt and Mani Madhukar, ISBN: 9781119592266. The book provides cutting-edge research that delivers insights into the tools, opportunities, novel strategies, techniques, and challenges for handling security issues in cloud computing, Internet of Things and social networking.

DESIGN AND ANALYSIS OF SECURITY PROTOCOLS FOR COMMUNICATION, Edited by Dinesh Goyal, S. Balamurugan, Sheng-Lung Peng and O.P. Verma, ISBN: 9781119555643. The book combines analysis and comparison of various security protocols such as HTTP, SMTP, RTP, RTCP, FTP, UDP for mobile or multimedia streaming security protocol.

SMART GRIDS FOR SMART CITIES VOLUME 2: Real-Time Applications in Smart Cities, Edited by O.V. Gnana Swathika, K. Karthikeyan, and Sanjeevikumar Padmanaban, ISBN: 9781394215874. Written and edited by a team of experts in the field, this second volume in a two-volume set focuses on an interdisciplinary perspective on the financial, environmental, and other benefits of smart grid technologies and solutions for smart cities.

SMART GRIDS AND INTERNET OF THINGS, Edited by Sanjeevikumar Padmanaban, Jens Bo Holm-Nielsen, Rajesh Kumar Dhanaraj, Malathy Sathyamoorthy, and Balamurugan Balusamy, ISBN: 9781119812449. Written and edited by a team of international professionals, this groundbreaking new volume covers the latest technologies in automation, tracking, energy distribution and consumption of Internet of Things (IoT) devices with smart grids.

inted and bound by CPI Group (UK) Ltd, Croydon, CR0 4YY
17/09/2026
14964247-0002